Onto-Ethologies

SUNY series in Environmental Philosophy and Ethics

J. Baird Callicott and John van Buren, editors

Onto-Ethologies

*The Animal Environments of
Uexküll, Heidegger, Merleau-Ponty,
and Deleuze*

Brett Buchanan

Published by
State University of New York Press, Albany

© 2008 State University of New York

All rights reserved

Printed in the United States of America

No part of this book may be used or reproduced in any manner whatsoever without written permission. No part of this book may be stored in a retrieval system or transmitted in any form or by any means including electronic, electrostatic, magnetic tape, mechanical, photocopying, recording, or otherwise without the prior permission in writing of the publisher.

For information, contact State University of New York Press, Albany, NY
www.sunypress.edu

Production by Ryan Morris
Marketing by Michael Campochiaro

Library of Congress Cataloging-in-Publication Data

Buchanan, Brett, 1975–
   Onto-ethologies : the animal environments of Uexküll, Heidegger, Merleau-Ponty, and Deleuze / Brett Buchanan.
      p. cm. — (SUNY series in environmental philosophy and ethics)
   Includes bibliographical references and index.
   ISBN 978-0-7914-7611-6 (hardcover) ISBN 978-0-7914-7612-3 (paperback)
   1. Philosophy of nature. 2. Naturalness (Environmental sciences).
3. Animals (Philosophy). I. Title.

BD581.B77 2008
113'.8—dc22
                                                  2008000121

10 9 8 7 6 5 4 3 2 1

Contents

Acknowledgments	vii
List of Abbreviations	ix
Introduction: Between Ontology and Ethology	1
1. Jakob von Uexküll's Theories of Life	7
Biography and Historical Background	9
Nature's Conformity with Plan	12
Umweltforschung	21
Biosemiotics	28
Concluding Remarks	36
2. Marking a Path into the Environments of Animals	39
The Essential Approach to the Animal	44
Heidegger and the Biologists	46
Three Paths to the World	55
3. Disruptive Behavior: Heidegger and the Captivated Animal	65
The Worldless Stone	69
The Poor Animal	71
Three Bees and a Lark	78
Animal Morphology	88
A Shocking Wealth	97
A Fine Line in the Rupture of Time	101
An Affected Body	112
4. The Theme of the Animal Melody: Merleau-Ponty and the *Umwelt*	115
The Structure of Behavior	116
A Pure Wake, A Quiet Force	130
A Leaf of Being	138
Interanimality	146

5. The-Animal-Stalks-at-Five-O'Clock: Deleuze's Affection
   for Uexküll   151
   Problematic Organisms   151
   Uexküll's Ethology of Affects   154
   The Body without Organs, the Embryonic Egg, and
      Prebiotic Soup   162
   Nature's Refrain Sung across Milieus and Territories   173
   The Animal Stalks   182

Conclusion: Uexküll and Us   187
Notes   191
Bibliography   207
Index   217

Acknowledgments

As this is a book on environments, I would be remiss if I did not say that it is very much a product of my own environment. Many people have read or listened to early drafts, and all have influenced it in one way or another. I wish to thank Will McNeill, Peter Steeves, and Tina Chanter, for their kindness and support, as well as Keith Bresnahan, Kathy Kiloh, Gerard Kuperus, Nat Leach, Leo Mos, Marjolein Oele, Tilottama Rajan, Nathan Ross, Dorian Stuber, Sokthan Yeng, and a whole host of friends and mentors at DePaul University, for their questions, encouragement, and friendship.

Laurentian University, and my colleagues especially, have been supportive in the completion of this project. Its completion has been aided, in part, through the Laurentian University Research Fund. This book has been strengthened by the insightful comments of two anonymous reviewers, and the helpful guidance of Ryan Morris, from the State University of New York Press. I would also like to thank John van Buren for his kind encourgement and advice from very early on, and Kenneth Paquette for his assistance on the manuscript.

My family has been supportive of my path into philosophy from the beginning and I owe them all inexpressible gratitude. Above all, a very special thanks to Kelly, the best part of my environment, and to whom this book is dedicated.

Abbreviations

Page numbers are to the original followed by the English (e.g., GA2, 49/75). In the case of Uexküll, page numbers are to the English translations. Translations have occasionally been modified.

WORKS BY DELEUZE

ATP — *Mille Plateaux* (avec Félix Guattari). Paris: Éditions de Minuit, 1980. (*A Thousand Plateaus* (with Félix Guattari), trans. Brian Massumi. Minneapolis, University of Minnesota Press, 1987.)

DR — *Différence et Répétition*. Paris: Presses Universitaires de France, 1968. (*Difference and Repetition*, trans. Paul Patton. New York: Columbia University Press, 1994.)

EPS — *Spinoza et le problème de l'expression*. Paris: Éditions de Minuit, 1968. (*Expressionism in Philosophy: Spinoza*, trans. Martin Joughin. New York: Zone Books, 1992.)

F — *Foucault*. Paris: Éditions de Minuit, 1986. (*Foucault*, trans. Seán Hand. Minneapolis: University of Minnesota Press, 1988.)

FB — *Francis Bacon: Logique de la Sensation*. Paris: Éditions du Seuil, 2002. (*Francis Bacon: The Logic of Sensation*, trans. Daniel W. Smith. Minneapolis: University of Minnesota Press, 2003.)

LS — *Logique du Sens*. Paris: Les Éditions de Minuit, 1969. (*The Logic of Sense*, ed. Constantin Boundas, trans. Mark Lester and Charles Stivale. New York: Columbia University Press, 1990.)

SPP — *Spinoza: Philosophie practique*. Paris: Éditions de Minuit, 1981. (*Spinoza: Practical Philosophy*, trans. Robert Hurley. San Francisco: City Lights Books, 1988.)

| WP | *Qu'est-ce que la philosophie?* (avec Félix Guattari). Paris: Éditions de Minuit, 1991. (*What Is Philosophy?* (with Félix Guattari), trans. Hugh Tomlinson and Graham Burchell. New York: Columbia University Press, 1994.) |

WORKS BY HEIDEGGER

BW	*Basic Writings*, ed. David Farrell Krell. New York: HarperCollins, 1993.
GA2	*Sein und Zeit*. Tübingen: Max Niemeyer Verlag, 2001. (*Being and Time*, trans. John Macquarrie and Edward Robinson. Oxford: Blackwell, 1962.)
GA8	*Was Heißt Denken?* Tübingen: Max Niemeyer Verlag, 1954. (*What Is Called Thinking?*, trans. J. Glenn Gray. New York: Harper & Row, 1968.)
GA9	*Wegmarken*. Frankfurt am Main: Vittorio Klostermann, 1976. (*Pathmarks*, ed. William McNeill. Cambridge: Cambridge University Press, 1998.)
GA12	*Unterwegs zur Sprache*. Pfullingen: Verlag Günther Neske, 1959. (*On the Way to Language*, trans. Peter D. Hertz. New York: HarperCollins, 1982.)
GA15	*Heraklit, Seminare* (with Eugen Fink). Frankfurt am Main: Vittorio Klostermann, 1970. (*Heraclitus Seminar*, trans. Charles H. Seibert. Evanston, IL: Northwestern University Press, 1993.)
GA20	*Prolegomena zur Geschichte des Zeitbegriffs*. Frankfurt am Main: Vittorio Klostermann, 1979. (*History of the Concept of Time: Prolegomena*, trans. Theodore Kisiel. Bloomington: Indiana University Press, 1992.)
GA21	*Logik: Die Frage nach der Wahrheit*. Frankfurt am Main: Vittorio Klostermann, 1976.
GA24	*Die Grundprobleme der Phänomenologie*. Frankfurt am Main: Vittorio Klostermann, 1975. (*The Basic Problems of Phenomenology*, trans. Albert Hofstadter. Bloomington: Indiana University Press, 1982.)
GA26	*Metaphysische Anfangsgründe der Logik im Ausgang von Leibniz*. Frankfurt am Main: Vittorio Klostermann, 1978. (*The Metaphysical

Foundations of Logic, trans. Michael Heim. Bloomington: Indiana University Press, 1984.)

GA29/30   *Die Grundbegriffe der Metaphysik. Welt-Endlichkeit-Einsamkeit.* Frankfurt am Main: Vittorio Klostermann, 1983. (*The Fundamental Concepts of Metaphysics: World, Finitude, Solitude*, trans. William McNeill and Nicholas Walker. Bloomington: Indiana University Press, 1995.)

GA31   *Vom Wesen der menschlichen Wahrheit.* Frankfurt am Main: Vittorio Klostermann, 1982. (*The Essence of Human Freedom: An Introduction to Philosophy*, trans. Ted Sadler. London: Continuum Press, 2005.)

GA33   *Aristoteles*, Metaphysik Θ *1–3: Von Wesen und Wirklichkeit der Kraft.* Frankfurt am Main: Vittorio Klostermann, 1990. (*Aristotle's Metaphysics* Θ *1–3: On the Essence and Actuality of Force*, trans. Walter Brogan and Peter Warnek. Bloomington: Indiana University Press, 1995.)

GA40   *Einführung in die Metaphysik.* Tübingen: Max Niemeyer Verlag, 1953. (*Introduction to Metaphysics*, trans. Gregory Fried and Richard Polt. New Haven: Yale University Press, 2000.)

GA42   *Schellings Abhandlung Über das Wesen der menschlichen Freiheit.* Tübingen: Max Niemeyer Verlag, 1971. (*Schelling's Treatise on the Essence of Human Freedom*, trans. Joan Stambaugh. Athens: Ohio University Press, 1985.)

GA54   *Parmenides.* Frankfurt am Main: Vittorio Klostermann, 1982. (*Parmenides*, trans. André Schuwer and Richard Rojcewicz. Bloomington: Indiana University Press, 1992.)

GA85   *Vom Wesen der Sprache. Die Metaphysik der Sprache und die Wesung des Wortes. Zu Herders Abhandlung "Über den Ursprung der Sprache."* Frankfurt am Main: Vittorio Klostermann, 1999. (*On the Essence of Language: The Metaphysics of Language and the Essencing of the Word. Concerning Herder's Treatise On the Origin of Language*, trans. Wanda Torres Gregory and Yvonne Unna. Albany: State University of New York Press, 2004.)

SEE   *Supplements: From the Earliest Essays to* Being and Time *and Beyond*, ed., John van Buren. Albany: State University of New York Press, 2002.

WORKS BY MERLEAU-PONTY

N	*La Nature: Notes Cours du Collège de France*, ét. Dominique Séglard. Paris: Seuil, 1995. (*Nature: Course Notes from the Collège de France*, trans. Robert Vallier.Evanston, IL: Northwestern University Press, 2003).

PhP	*Phénoménologie de la perception*. Paris: Gallimard, 1945. (*The Phenomenology of Perception*, trans. Colin Smith. New York: Routledge, 1998).

S	*Signes*. Paris: Gallimard, 1960. (*Signs*, trans. Richard C. McCleary. Evanston, IL: Northwestern University Press, 1964).

SB	*La structure de la comportement*. Paris: Presses Universitaires de France, 1943. (*The Structure of Behavior*, trans. Alden Fischer. Pittsburgh: Duquesne University Press, 1983).

VI	*Le visible et l'invisible*. Paris: Gallimard, 1963. (*The Visible and the Invisible*, trans. Alphonso Lingis. Evanston, IL: Northwestern University Press, 1968.)

WORKS BY UEXKÜLL

IU	"An introduction to Umwelt," trans. Gösta Brunow. *Semiotica* 134, 1/4 (2001): 107–110.

NCU	"The new concept of Umwelt: A link between science and the humanities," trans. Gösta Brunow. *Semiotica* 134, 1/4 (2001): 111–123.

SAM	*Streifzüge durch die Umwelten von Tieren und Menschen* (with Georg Kriszat). Geibungsyoin Verlag, 1934. ("A Stroll Through the Worlds of Animals and Men," *Instinctive Behavior: The Development of a Modern Concept*, ed. and trans. Claire Schiller. New York: International Universities Press, 1957.)

TB	*Theoretische Biologie*. Berlin: J. Springer Verlag, 1926. (*Theoretical Biology*, trans. D. L. Mackinnon. New York: Harcourt, Brace, 1926.)

TM	*Bedeutungslehre*. Leipzig: Verlag von J. A. Barth, 1940. ("The theory of meaning," *Semiotica* 42, 1 (1982): 25–82.)

UIT	*Umwelt und Innenwelt der Tiere*. Berlin: J. Springer Verlag, 1921. ("Environment and Inner World of Animals," excerpts trans. Chauncey J. Mellor and Doris Gove. *Foundations of Comparative Ethology*, ed. Gordon M. Burghardt. New York: Van Nostrand, 1985).

Introduction

Between Ontology and Ethology

In 1934, the biologist Jakob von Uexküll (1864–1944) wrote a small "picture book" that he whimsically titled *A Stroll Through the Environments of Animals and Humans*. While the title certainly captures the casual attitude that pervades this monograph, it belies the more radical venture that Uexküll presents in his theorization of animal life. He is interested in what it is like to be an animal and, as we shall see, it has everything to do with the reality of the environment. This is not Uexküll's first attempt at articulating the meaning of the environment beyond a strictly human perspective, but it is by far his most popular to date. In the foreword, Uexküll appropriately sets the stage:

> The best way to begin this stroll is to set out on a sunny day through a flower-strewn meadow that is humming with insects and fluttering with butterflies, and build around every animal a soap bubble [*Seifenblase*] to represent its own environment [*Umwelt*] that is filled with the perceptions accessible to that subject alone. As soon as we ourselves step into one of these bubbles, the surrounding meadow [*Umgebung*] is completely transformed. Many of its colorful features disappear, others no longer belong together, new relationships are created. A new world emerges in each bubble. The reader is invited to traverse these worlds with us. (SAM, 5)

The invitation is deceptively simple: he claims that we are not heading toward a new science, but merely strolling into unfamiliar, invisible, and previously unknown worlds. With Uexküll leading the way, we are not undertaking the usual Sunday stroll. No, it may not be a new science, not nearly so ordinary and pedantic, but it is indeed something wondrous. New worlds arise before our eyes, through our sensations, in our imaginations. We are asked to step out of ourselves and into the strange environments of bees, sea anemones, dogs, ticks, bears, and many others. Uexküll's illustrator, Georg Kriszat, even provides some illustrations to help us make the leap.

What concerns Uexküll here, as well as elsewhere in his writings, is how we can glimpse natural environments as meaningful to the animals themselves. Rather than conceiving of the world according to the parameters of our own human understanding—which, historically, has been the more prevalent approach—Uexküll asks us to rethink how we view the reality of the world as well as what it means to be an animal. So not only does he multiply the world into infinite animal environments, he also seeks to transform our understanding of the animal away from its traditional interpretation as a soulless machine, vacuous object, or dispassionate brute. Against such positions, Uexküll proposes to understand the "life story" of each animal according to its own perceptions and actions: "We no longer regard animals as mere objects, but as subjects whose essential activity consists of perceiving and acting. We thus unlock the gates that lead to other realms, for all that a subject perceives becomes his perceptual world [*Merkwelt*] and all that he does, his active world [*Wirkwelt*]. Perceptual and active worlds together form a closed unit, the *Umwelt*" (6). As one can see, there is a good deal of secrecy and mystery involved in this project. Uexküll wishes to unlock the gates into previously forbidden worlds, all the while retaining the closed bubble intact. His focus is on the subjective animal and its unique environment, each giving rise to new relationships that were previously unappreciated. As with all things otherworldly, this invitation involves a bit of speculation.

One of the conclusions that he reaches is that insofar as each animal constructs its own environment out of the midst of its perceptions, actions, and relationships, "there are, then, purely subjective realities in the *Umwelten*; and even the things that exist objectively in the surroundings never appear there as such" (72). Part of his appeal—aside from introducing even the most skeptical reader to look at animals and the environment differently—is that he is on the verge of producing an ontology of the animal from his ethological observations. "Ethology" will not become a common concept until Konrad Lorenz later popularizes it, but Uexküll's biology, as Lorenz himself notes, is a pioneering study of animal behavior. But it is not only that. Perhaps it was his lingering interest in philosophical ideas, perhaps it was just a result of his strolls; either way you look at it, Uexküll also introduces us to a new way of thinking about reality as such. He is not the first, of course, to suggest that reality is more than one physical world, but he is one of the first to really push the subjective experience of the animal. The being of the animal unfolds through its behavior and we catch a glimpse of this from within its own unique environment, its bubble-like *Umwelten*.

From out of this Uexküll leaves open an interesting question: what is the role of the body? Toward the end of his analysis we are told "many problems await conceptual formulation, while others have not yet developed beyond the stage of formulating questions. Thus we know nothing so far of the extent to which the subject's own body enters into his *Umwelt*" (73). That the question of the body has been left unanswered is particularly intriguing given that, within the tradition of Western philosophy, it is almost always the case that the body is that which is specifically animalistic. Where mind, consciousness, rational thought, and spirit have been affiliated with human life, the opposite has also been true: the body is that which is instinctual, sensual, mechanical, finite—and animal. It is striking, then, that Uexküll claims that the animal's body has not yet received adequate attention within his discussion of animal environments, for where else has the subjective reality occurred if not in and through the behavioral body? How does the body enter into relation with, and creation of, the environment? Similarly, how does the environment enter into the body of the animal? How does such a relationship between body and environment force us to rethink both concepts? It is within this interaction between body and environment that animal behavior reveals its ontological dimensions.

It may very well be that Uexküll opens many more questions than he is able to answer. Yet it is clear that the questions emerging from his project have not gone unheard. Despite his relative anonymity, many have quietly taken up the loose threads of his thought and applied it to studies ranging from classic ethology to cognitive neuroscience and from linguistics to art and philosophy. Not all have turned to him for the same reason, but each has found in Uexküll's thought something compelling, whether it is a lingering problem or hidden insight. Within continental philosophy alone, he has appeared in many of the more formative thinkers of the twentieth century, including Martin Heidegger, Ernst Cassirer, Hans-Georg Gadamer, José Ortega y Gasset, Jacques Lacan, Maurice Merleau-Ponty, Georges Canguilhem, Gilles Deleuze, and Giorgio Agamben.[1] It would be fair to say that many have taken Uexküll's stroll.

In calling the present book "Onto-ethologies," I am consciously situating this work within a growing field that brings continental philosophy together with the sciences. Granted, this is a large endeavor, and one that extends far beyond the pages committed here. However, to reiterate the telling words of Keith Ansell Pearson, biophilosophy has been a relatively neglected tradition in contemporary thought. This is not to suggest that this field is a new one. This would be grossly unfair to the many who have staked their territory in this very area, and have done so from fairly early on, such as Marjorie Grene, Hans Jonas, and many others, not to mention a whole French tradition of the life sciences that, as Michel Foucault has remarked, remained overshadowed by the dominance of phenomenology.[2] This is also to say nothing of the close relation between the sciences and philosophy in centuries past. But whereas continental philosophy is better known for its engagement with the history of philosophy, the arts, ethics, politics, and its critiques of metaphysics, it is only recently that a more concerted emphasis has been placed again on its diverse relations with the sciences. As an example, we can observe the proliferation in recent years of studies bearing on biophilosophy, zoontology, geophenomenology, geophilosophy, ecophenomenology, animal others, and many others, to say nothing of the growing field of animal studies itself.[3] All of these studies point more toward a significant direction in our theoretical framework than to some transitional fad. While we emerge from the critiques of our humanist tradition, an increasing focus is being paid to how and where we find ourselves within nature and the world at large. "What does it mean to be human?" becomes a question of life and the living being. Attention to the status of human beings need not disappear, of course, but it does become framed by a broader emphasis on nature and life. What are the relations within nature? And how do we assess these ontological relations?

Within this midst, I submit my contribution on onto-ethology.[4] The aspiration I have with this study is to examine how three philosophers (Heidegger, Merleau-Ponty, and Deleuze) each entertain the thought of a single biologist (Uexküll). As noted, many have approached Uexküll's thought, but these three in particular have found in him a compelling case for an ontology of living beings. Similarly, all three have approached different biologists over the course of their writings, but it is with Uexküll's ethology that one observes a lasting impression. In every case, ethology emerges as the significant dimension in framing the being and becoming of the animal. The animal body is interrelated with its environment through the process of behavior, so it becomes a question of how to engage the ontological dimension of this relation. Each therefore discovers something different in Uexküll as they reveal the wonders that come from relearning to look at animal environments.

The first chapter presents a general overview of Uexküll's research program. Beginning with a brief biographical and historical introduction to the climate of his biological studies, I will then turn to three areas of his theoretical biology. First, Uexküll's theory of life falls within a broad conception of nature that holds all of life as conforming to a plan. His early studies in biology lead him to believe that even if nature is neither teleological nor mechanical, both dominant and competing perspectives on nature at the time, it nevertheless follows a plan. Examples of this derive from his embryological and physiological studies, but Uexküll sees evidence of a plan in how all of nature coheres together like a great symphony. All the various parts appear to work in harmony with one another, as found, more specifically, in the relations animals have with their environments. Coinciding with his *Umwelt* research, Uexküll is also considered a pioneer in biosemiotics, the field of study that examines how sign systems are produced and interpreted within nature. By preparing the basis for a theory of biological signs, Uexküll becomes his most adventuresome: nature as the embodiment of significance and the possibility of meaning in life. Some of the characters of his story will include the tick, spider, fly, bee, flower, and the unnamed mammal.

The second chapter begins to set the stage for the approach to the ontological problem of the environment and world. Nowhere is this more evident than Heidegger's early investigations into how the world is the meaningful horizon for human existence. It is in this context that the question of the animal is first posed. If human Dasein is defined ontologically as being-in-the-world, what is the relation that animals have to the world? Does the world reveal itself to animals as it does within human existence? In order to arrive at this scenario, Heidegger's treatment of biology—and his impressions of notable biologists—is first presented as a means to arrive at how and why the world is an important factor in understanding the being of animals. Uexküll's early writings on animal *Umwelten* emerge as a favorable basis to compare his own understanding. The world is thus treated according to its philosophical etymology, according to the everyday understanding of being-in-the-world, and finally as a comparative analysis between human existence and animal life.

The third chapter looks to Heidegger's writings on animal life, in which Uexküll figures as representative of both the best and most disappointing in contemporary biology. Heidegger postulates a trio of theses with respect to beings and the world: the stone is worldless, the animal poor in world, and the human world-forming. In the end, we learn that animals do not exist, so to speak, insofar as they are unable to transcend their captivation by things. Animals admittedly have relations with things in their midst, and they do so through the outward extension of their body, but they are said to lack an access to the things in themselves, to the being of these beings. In this view, Heidegger postulates animal behavior as restrictive in comparison to the comportment of human Dasein, which is unbound and free in its opening of the world. Animal behavior thus indicates an inability to transcend the peculiar manner of animal being and so too an incapacity of the animal's ever relating to an environment. But the body and environment remain as ontological problems for Heidegger, and Uexküll's animals persist throughout his career.

The fourth chapter begins with Merleau-Ponty's early writings on behavior. His interest is primarily in how the relation between consciousness and the natural world can suggest a means of overcoming an overly materialist interpretation of the world. Most noteworthy, however, are his reflections on how behavior, far from coinciding with the mechanist argument of cause and effect, actually serves to provide a glimpse of the animal as a totality greater than the sum of its parts. Taking his cue from Uexküll, Merleau-Ponty depicts the being of the animal as part of a melodic and rhythmic order whereby the world opens through behavior itself. Merleau-Ponty returns to Uexküll in his late writings on nature while he is in the process of reconsidering nature from an ontological point of view in his theories of the "flesh." Of particular interest are the themes of the animal melody, interanimality, and how the animal body inscribes an intercorporeal *Umwelt* through behavior. The animal remains a consistent theme over the course of Merleau-Ponty's career, and serves an important role in the development of his own ontology.

The fifth chapter provides a look at how Deleuze, both in his individual writings and in his collaborations with Félix Guattari, reads Uexküll as a Spinozian ethologist. Deleuze's ontology is imbued with references to biological life, from *Difference and Repetition* to his final works. But rather than concerning himself with the animal–environment relation, he is more interested in the virtual and intensive processes that create actual beings and their relations. In Uexküll he discovers an ethologist who counts the affects of bodies. Deleuze therefore provides a startlingly creative way of reading Uexküll such that many, if not most, of the previous concepts (body, environment, behavior, organism, animal) are subjected to reconsideration. His ontology depicts a new way of seeing animal life as a continual process of becoming, where bodies are always changing, and entities like the organism represent not life but life's imprisonment. Ethology becomes a study of counting affects, and the result is a shifting landscape of bodies forming new patterns in the midst of nature.

CHAPTER 1

Jakob von Uexküll's Theories of Life

In 1952, Georges Canguilhem, the great historian and philosopher of the sciences, remarked that the concept of the environment (*milieu*) was becoming indispensable in the consideration of living beings. In *La connaisance de la vie*, he writes: "The notion of the milieu is in the process of becoming a universal and obligatory mode to capture the experience and existence of living beings. We can almost even say that it forms a necessary category of contemporary thought" (129). This is quite the claim, particularly since it was not always so. For quite a while, the living being was conceptually displaced from its natural milieu. Though Uexküll figures as only part of Canguilhem's historical account, he was nevertheless a key facilitator in this contemporary focus on animal environments. From as early as 1909 with the publication of *Umwelt und Innenwelt der Tiere*, through to the end of his life in 1944, Uexküll focused his research on attempting to discern and give expression to the "phenomenal worlds" (SAM, 7) and "subjective universes" (TM, 29) of animals. Each of these terms, however, is just a different way of translating Uexküll's new concept of "*Umwelt*," a term that more literally means "surrounding world" or "environment," but that I will retain in the original language.[1] His contention was that conventional biology had run its course by treating animals as objects governed by mechanical laws of nature such that they became accessible to the scientific eye of human objectivity. If biology continued to understand animal life with misguided objectivity, it would eventually succumb to the influence of chemistry and physics by seeking, wrongly in his estimation, to ground its knowledge in the reductionist accounts of chemico-physical factors. Much of his treatise on *Theoretical Biology* (1920) explicitly attends to the differences between biological thought and the seemingly wayward ways of physics and chemistry. Rather than continuing to understand animals as "physico-chemical machines" (TB, xiii), Uexküll contends that animals must be interpreted by virtue of the environments that they inhabit, and, insofar as it is possible, from the perspective of their behavior within such environments. The biologist must do so, moreover, while remaining free from the inclination to anthropomorphize the *Umwelten* of animals and, as Marjorie Grene has noted, retaining the rigorous accuracy expected from science.

These observations lead us to discern a number of key aspects that Uexküll introduces with his *Umwelt* research. In order to give a brief indication of the direction I intend to take in the ensuing pages, the following can be said concerning Uexküll's research. Uexküll firmly believes that nature conforms to a plan (*Planmäßigkeit*) whereby organic and inorganic things cohere together in great compositional harmony. The musical reference is a consistent one in his literature and is crucial to understanding how he interprets organisms as 'tones' that resonate and harmonize with other things, both living and nonliving. Nature's conformity with plan is based partially in Kantian and Baerian terms; I will explore both of these bases. The melodic perspective also leads Uexküll to differentiate himself from Darwin's theory of evolution, which he saw as a 'vertical' model of descent and one that emphasizes far too much a chaotic view of nature's formations. Uexküll was not necessarily anti-evolutionary, but his focus was certainly directed elsewhere, specifically toward a more 'horizontal' model that looks at how organisms behave and relate to things across their respective environments. Instead of interpreting organisms based on natural selection, for instance, Uexküll sought to understand them with respect to the designs that they represented in relation to meaningful signs. This led his research toward positing an ethological study of animal behavioral patterns, anticipating the work of such notable ethologists as Konrad Lorenz and Niko Tinbergen. *Umwelt* research also led him to be an early pioneer of a field that would become known as biosemiotics. In studying the behavioral patterns of different animals, Uexküll noted that animals of all levels, from microorganisms to human animals, are capable of discerning meaning from environmental cues beyond a purely instinctual reaction. Such meaning is attributable to how organisms enter into relationships with other things and thus come to see the environment as laced not just with signs, but with significance itself. The nature of these relations, and more specifically how one interprets them, will have profound consequences when it comes to discerning certain differences between Heidegger's, Merleau-Ponty's, and Deleuze's ontologies.

To better explore these themes, this chapter is divided into the following sections: (1) a brief biography and historical background to Uexküll's biology, (2) nature as conformity with plan, (3) *Umwelt* research, and (4) biosemiotics. Each section is aimed at being faithful to Uexküll's thought while at the same time anticipating the philosophical readings of Heidegger, Merleau-Ponty, and Deleuze and Guattari.

BIOGRAPHY AND HISTORICAL BACKGROUND

Jakob von Uexküll was born in Keblas, Estonia, in 1864, to parents of modest means. His father had interests in politics and became mayor for a short period of the small town of Reval. Uexküll studied zoology from 1884 to 1889 at the University of Dorpat (now the University of Tartu) where he was unquestionably influenced by two strong and contrasting schools of biological thought: the emergence of Charles Darwin's theories (1809–1882) and the legacy of Karl Ernst von Baer (1792–1876). As one commentator explains, one of Uexküll's professors at Dorpat was Georg Seidlitz (1840–1917), a Darwinian scholar who is held to be one of the first to teach Darwin's theory of evolution within continental Europe.[2] It is unclear just how much Seidlitz influenced Uexküll's studies, but as we will see, Uexküll was in the end not very convinced by Darwin's theory of evolution. This may be due to the other and more dominant school of thought at Dorpat, where the influence of Baer, who was himself educated at Dorpat, left a strong presence within the zoology department even after his death.

The schism between the Baerian and the Darwinian influences is fairly representative of a general tension in nineteenth-century German biology. Biology itself, as a formal and unique science, wasn't actually coined until 1802, when both Jean-Baptiste Lamarck and Gottfried Reinhold Treviranus coincidentally first used the term.[3] From its onset, the debate in biology during this period surrounded the issue of how and whether one could understand natural life in a manner equal to Newton's discoveries in physics. In part, biological thought was immediately immersed in the problem of either reconciling or favoring one of two views: the teleological view of nature that found its roots in Aristotelian science and a mechanistic science that found nature obeying unwavering physical laws. Both trends—teleology's necessary goal-directedness and mechanism's lawful accidents—likewise found a philosophical impetus in the works of Immanuel Kant and, following him, in the *Naturphilosophie* of G. W. F. Hegel and Friedrich Schelling. This dichotomy between teleology and mechanism had many voices on either side, but, for our purposes, it suffices to mention that two of the major proponents in biology included Baer's teleological view and Darwin's mechanist theory.

During his academic education, Baer was taught by the biologist Ignaz Döllinger. Döllinger was a close adherent of Schelling's philosophy of nature, and he had also studied under Kant for a short period in Königsberg. This coupling of biology with a philosophy of nature trickled through Döllinger into the works of Baer; however, it was Georges Cuvier and Kant who had the greatest impact on Baer. It is also notable that Baer's first academic posting was in Kant's hometown of Königsberg; even though he didn't teach there until after Kant's death, the connection between Baer and Kant's thought was already secured through his education.

Baer's focus in biology was in the emerging field of embryonic morphology, the study of embryonic forms, and, more specifically, *Entwicklungsgeschichte*, the developmental theory of animal organization. Baer believed that the embryos of all organisms have a purposefulness (*Zielstrebigkeit*) in the unfolding of their development. Each part or organ of the embryo develops according to a plan that demonstrates the overall organization of each organism. Baer outlined four rules over the course of his observations, and all four have come to be summarily known as "Baer's Law," which states that the development of the embryo moves from very general characteristics to more particular and specific ones.[4] Baer's studies are important for many reasons, not the least of which is his strong contribution to the epigenetic theory of embryonic development in contrast to the increasing skepticism surrounding the theory of preformationism, which holds that embryos are already 'preformed' organisms from conception. Baer's argument that observational studies of embryos demonstrate a movement from an indistinct and general form toward an increasingly specific form was quite significant. However, Baer does not jump to the conclusion that all organisms must descend from the same origin, as though all species descended from a primal *Ur*-organism. Rather, all organisms are said to belong to four "types," each of which manifests its own distinctions in morphology, and each therefore has its own general characteristics.

These studies have also ensured Baer a place within the teleological camp. But his teleology is not one that assumes a cosmological aim toward which all of nature is heading, nor does it make a claim for a rational mind or God behind the developmental process. Instead, Baer's teleology is what Timothy Lenoir describes as a "vital materialism," whereby all of nature's entities have "an emergent property dependent upon the specific order and arrangement of the components" (9). Each organism, in other words, develops according to a plan, leading from the general characteristics of its type to the particular traits of that specific organism.

This position will eventually put Baer's teleological view in an irreconcilable position with Darwin's theory of evolution. The difference will be formulated, however, not at the level of morphology, but in the mechanism behind Darwin's theory of evolution. It is well known that when Darwin published *The Origin of Species* in 1859, the most radical idea wasn't evolution itself (the general idea that had been floating around for some time), but the mechanism behind evolution, namely, natural selection. Natural selection, as Daniel Dennett accurately describes, was "Darwin's dangerous idea" because it "unifies the realm of life, meaning, and purpose with the realm of space and time, cause and effect, mechanism and physical law" (21). Darwin, in a word, accounts for the unfolding of species not according to any specific plan or goal, but through a war of attrition where the weak are weeded out, the strong survive, and, more important, pass on their genes to later generations (though Darwin himself could not prove how this last genetic step worked). Natural selection is a dangerous idea for many reasons, perhaps the greatest of which is its ability to offer an observable, testable, and scientific account for evolution, where the repercussions extend into philosophical and religious beliefs. However, what Baer responded to in his manuscript *Über Darwins Lehre* in 1873 was the seemingly accidental and planless nature of Darwin's theory of evolution. As Lenoir claims, Baer was less concerned with denying evolution as such than with offering a "theory of limited evolution" confined to demonstrating a "parallel between the general pattern of ontogenesis and organic evolution" (264–65). This could also explain why Baer entitles his book the way he does: it is a treatise *Über Darwins Lehre* (On Darwin's Theory) rather than *Gegen Darwins Lehre* (Against Darwin's Theory). Evolution, for Baer, is a phenomenon best described in terms of development. Stephen Jay Gould explains this point: "Evolution occurs when ontogeny is altered in one of two ways: when new characters are introduced at any stage of development with varying effects upon subsequent stages, or when characters already present undergo changes in developmental timing" (4). What threatened Baer was the unaccountable phenomenon of natural selection that seemed to overrule the orderly and directed development of organisms. Baer's dispute with Darwinian evolution was therefore oriented toward saving a teleological view of morphology against the overly mechanical and seemingly accidental view of development offered by Darwin.

Such was the intellectual situation in biology when Uexküll studied at Dorpat. The debate between teleological and mechanistic interpretations of natural life was far from over, and even continues to this day, so it is no surprise that it had a decisive influence on the young Uexküll. As will be seen, Uexküll was particularly averse toward the Darwinian theory of evolution, and he was so in a manner peculiar to the formulation of his own developing thought. However, even though Uexküll continued to find himself siding with the historically less popular Baerian interpretation of biology, we cannot forget or dismiss as merely coincidental that he studied biology at Dorpat, Baer's alma mater, just eight years after Baer's death.

After his undergraduate education, Uexküll went on to complete his studies at the University of Heidelberg, where he worked in the field of muscular physiology, particularly of marine invertebrates. He studied under the directorship of Wilhelm Kühne (1837–1900), whom he had met in Dorpat on the occasion of a memorial for Baer's death in 1886. After receiving an honorary doctorate at Heidelberg in 1907, Uexküll worked at the Zoological Center in Naples before eventually founding the Institute for Umwelt-Research in 1926 at the University of Hamburg. While he finished his career in Germany, Italy proved to be his true love and final residence. As Giorgio Agamben suggests, Uexküll had to leave the southern sun of Italy due to the dwindling finances of his familial inheritance, but he still kept a villa in Capri, to which he would occasionally return and eventually spend the last four years of his life. It is also suggested that Walter Benjamin, the German Jewish critical theorist, stayed for several months at Uexküll's villa in 1924. While this encounter probably had little effect on either's work, it is nevertheless interesting in situating Uexküll within the parameters of this intellectual history. The final years of his life were spent with his wife—who would write his biography a decade after his death—in Capri.

Over the course of his life, Uexküll wrote well over a dozen books, as well as many more scientific articles, covering a wide range of topics from the physiological musculature of marine invertebrates to the subjective lives of animals, from God and the meaning of life to biological readings of Plato and Kant. Among the most influential of his works are the aforementioned *Umwelt und Innenwelt der Tiere* (1909), *Theoretische Biologie* (1920), *Die Lebenslehre* (1930), *Streifzüge durch die Umwelten von Tieren und Menschen* (1934), *Niegeschaute Welten* (1936), and *Bedeutungslehre* (1940).

NATURE'S CONFORMITY WITH PLAN

If biology, as Uexküll understands it, is the "theory of life," then one might best begin by asking what life is in order to arrive at his biology. Toward the end of his life, Uexküll will place more and more emphasis on "meaning" and "significance," stating in *The Theory of Meaning* "that life can only be understood when one has acknowledged the importance of meaning" (26). But before addressing the theme of meaning in the section on biosemiotics, we can observe how Uexküll eventually comes to focus on meaning and signification via his early theory on nature's conformity with plan (*Planmäßigkeit*). In fact, one can read the development of his thought as leading from theoretical biology to a general concept of life as inherently meaningful, as I will propose here. Nature's conformity, as he states in *Theoretical Biology*, "is the basis of life" (xi), so we turn first to this before turning our attention to how life might be thought of as meaningful.

Uexküll opens his largest and most comprehensive text, *Theoretical Biology*, with an acknowledgment to an unlikely source: the German philosopher Immanuel Kant (1724–1804). In his introduction, Uexküll writes: "The task of biology consists in expanding in two directions the results of Kant's investigations:—(1) by considering the part played by our body, and especially by our sense-organs and central nervous system, and (2) by studying the relations of other subjects (animals) to objects" (xv). Before examining these two points in further detail, we need to know what exactly Uexküll means by "the results of Kant's investigations," such that we understand his biology as expanding on it. To do so, one need only look prior to this enumeration, where he offers a rather succinct, though largely undeveloped, interpretation of Kant's philosophy, when he states that "*all reality is subjective appearance* [Alle Wirklichkeit ist subjective Erscheinung]" (xv).

Uexküll takes as his guiding philosophy a thesis that will provide the foundation for the entirety of his thought: that the reality we know and experience is ultimately what we subjectively perceive in the world. There is no objective reality in the form of objects, things, or the world; there is nothing outside of the individually subjective experiences that create a world as meaningful. If Uexküll has a biological ontology, it is here. He will add layers and depth to this position, but the foundation is already set. Reality is created through the experiences of each and every subject, and this, as we shall see, holds for all animals just as much as it does for humans. Uexküll is clearly inspired by Kant's self-proclaimed second Copernican revolution; this is Kant in his most familiar form. In the preface to the second edition of *Critique of Pure Reason*, Kant writes of the "altered method of our way of thinking, namely that we can cognize of things a priori only what we have put into them" (Bxviii; Bxxii). By likening his thought to Copernicus, Kant sought to reevaluate the role that the perceiver plays in knowing the surrounding world of things. Instead of assuming an objective world that exists independent of the subjective perceiver, Kant reformulated the question by asking whether it may not be we who are subjectively, albeit a priori, forming our knowledge of the world. It is no longer thought that our ideas and thoughts mirror the world outside us, but that the world conforms to our cognitive faculties. If this is the case, then it remains the task of the philosopher to ascertain the categories of the mind that allow for our sensibility and understanding to construct such a world in which we live. Alas, this remains the critique of pure reason and not the task of theoretical biology. For our purposes, let it suffice to note that Uexküll more or less takes Kant at his word by glossing over his position, and thus concludes that "Kant had already shaken the complacent position of the universe by exposing it as being merely a human form of perception" (IU, 109). Uexküll expands this thought, however, by attributing subjective perception to not just human forms of perception but to the *Umwelten* of all animal perceptions.

What is further noteworthy in Uexküll's adoption of the subjective position is that he repudiates the notion that we will ever get to a reality outside of subjective perceptions. On this point he differs from Hermann von Helmholtz, whose work he often cites as informative to his own observations, and it could be for this reason that he makes an appeal to Kant's philosophy. Still within the introduction to *Theoretical Biology*, it is admitted that "Helmholtz indeed acknowledged that all objects must appear different to each subject; but he was seeking the reality behind appearances" (xv). It is possible then that Uexküll took Helmholtz as his starting point for observing the subjective appearance of reality, but that he found Helmholtz overextending himself into an area that he ought to have left well alone. Like Kant, Uexküll did not believe that we could get to a noumenal "thing in itself"; all that we have are phenomenal appearances. Helmholtz believed this reality behind appearances to be "the physical laws of the universe," but, for Uexküll, such a reality can only be tenable as an article of faith, not of science.

This prepares the way for Uexküll's rejection of certain physical principles on the basis that he finds biology to be largely nonmechanistic. A significant theme of his theoretical biology is to underscore, in a decisive manner, how and why biology is different from the other natural sciences, specifically physics and chemistry. This includes, among other things, claiming how organisms are different from machines, which he answers in a twofold fashion: by referring to living things as both self-developmental and autonomous. With the first, Uexküll contrasts the "centripetal architecture" of purely physical things with the "centrifugal architecture" of organisms; the former accounts for how material things are formed by outside forces acting inwardly, whereas with the latter we are led to see how organisms develop from the inside out (TB, 190). This highlights the role that Uexküll gives to morphology. Living things develop, from the blastula phase on, in a coherent, self-regulated way directed by inner principles. This importantly does not preclude outside agents acting on the genesis of the living thing. It will be quite the contrary, as we will see in his descriptions of the *Umwelt*. So while the contrast with machines is perhaps simplistic, the point he makes is clear: material, nonliving things are created from the outside by parts being put together or taken apart, whereas living, organic beings develop from an inner force that unfolds according to a morphological plan. Living things are always already a completed unity, no matter what stage of development, in a way that objects and machines cannot be. The vital materialism of Baer's morphological studies is evident here in Uexküll's account, though I'm not sure we can go so far as to call him, as Lorenz lovingly does, a "dyed-in-the-wool vitalist."[5] The inner force, as we shall see, is offset by environmental factors.

The centrifugal theory of development coincides nicely with his second point, notably the claim that living things are autonomous beings not dictated by physical laws alone. One of the central features that distinguishes the living from the nonliving is that living things are subject to their own self-governing laws. According to Uexküll, "to be a subject means, namely, the continuous control of a framework by an autonomous rule" (TB, 223). His use of autonomy is fairly literal. It is not an issue of an organism's freedom to do what it wants, but its natural inclination to self-rule. It abides by its own principles, no matter how fixed these may be, and not the rules of another. Organisms, therefore, are not mere machines because of their inner morphological development and of their autonomy. They are understood as a whole, not by divisible parts.

The point here is that Uexküll does not believe biology ought to inquire into the domain of physics and chemistry, for to do so leads toward positing absolute laws, such as Helmholtz's "physical laws of the universe." Nor should physics and chemistry intrude on biology. To do so would require formulating problems and answers irrespective of the uniqueness of the living being in question. This marks a significant departure in theoretical biology, for Uexküll believes that biological thought has been under the influence of the chemical and physical sciences for too long. His claim is all the more provocative due to its parallel with the contemporaneous critique of metaphysics present during his time. It is interesting to read that the belief in an objective reality underlying the apparent world has not only been a thorn in the side of postmetaphysical thought, as found with Nietzsche for instance, but that this belief has also undermined the advancement of biology because of physics' proximity with such a metaphysics. In fact, Uexküll makes the startling claim that "present-day physics is, next to theology, the purest metaphysics" (TM, 42) precisely because of its faith in an idealized objective world that presumably lies beyond the temporary fleetingness of subjective appearances. Biology is simply not in the same company as (meta)physics: "Biology does not claim to be such extensive metaphysics. It only seeks to point to those factors present in the living subject that allow him to perceive a world around him, and serve to make this world of the senses coherent" (TM, 43). To this end, "it seems," Uexküll notes, "that we must abandon our fond belief in an absolute, material world, with its eternal natural laws, and admit that it is the laws of our subject" that constitute the world as meaningful (TB, 89).

The path that Uexküll is navigating is a difficult one. On the one hand, he finds impetus for the future of biological thought in the guidance of Kant's philosophy. Here we find, in Uexküll's reading, that there is no truly objective world other than what we subjectively perceive. That which is known cannot exceed an irreducible world of experience; there can be no absolute world from a biological position. He maintains this to be true of the entire natural world, from the simplest to the most complex of organisms. In making this claim, he likewise shies away from a world of pure causality where everything can be explained by mechanical and physical laws. On this point, Uexküll is clear that we must distance biology from physics if we want to address nature's plan. As he explains, "[p]hysics maintains that the things of Nature around us obey causality alone. We have called such causally ordered things 'objects.' In contrast to this, biology declares that, in addition to causality, there is a second, subjective rule whereby we systematize objects: this is conformity with plan, and it is necessary if the world-picture is to be complete" (TB, 103). For him, "*all reality is subjective appearance*" because reality is constituted by living things that are subjects themselves, even if they together constitute a greater plan. On the other hand, however, Uexküll does not want biology to devolve into an entirely relativist science, where the world can be interpreted any which way and where nature is subject to a variety of accidental, random, and chaotic events. For better or worse, this is his impression of Darwinism and the theory of evolution more specifically. In order to further clarify this thought, his "conformity with plan" must therefore be situated between the too-strict objectivity of physical mechanism and the too-random planlessness of Darwinism.

Uexküll most clearly distances his theory from Darwinism by offering a brief narrative of the history of science from Kepler to Darwin. In both his *Theoretical Biology* and, in greater detail, "The new concept of Umwelt" (1937), Uexküll describes how science passed, between the time of Kepler and Newton, from a "*perceptual*" orientation to a "*functional*" view of the universe. The description is as literal as it sounds: modern science originally arose through the observation of natural things, from plants and animals to the distant stars in the sky above. Such perceptual observations, however, gave way to a more rigorous study of how things in the universe function independent of the observer. For example, he suggests that modern astronomy originated as a perceptual study of heavenly bodies by wondering about the likelihood of a design behind their observable movements. The harmony of these movements was attributed to God who alone was thought capable of ordaining such a perfect cosmic balance. However, with the emergence of Newton's natural laws, Uexküll finds that perceptual study succumbed to a study of function to such an extent that "causation" came to overrule "design" as the guiding principle of science. Newton's discoveries had such an impact that the physical and chemical sciences "busied themselves with the functional side of things and shoved the perceptual side away with scorn. Both acknowledged only the law of cause and effect and denied the existence of design in nature" (NCU, 114). Together with this shift, "something fundamentally shattering had happened—God had left the universe."

The absence of God is important here not only because Uexküll finds the death of God in Newtonian science (though this is certainly interesting), but because his argument against causation and function hinges on the relation to God, more so than on perceptual science. Studying the world as if it were a mechanical machine seems to imply that design is no longer possible and that God is no longer necessary. Without God as the designer, the world becomes a machine simply going through mechanical and ultimately meaningless motions. Unfortunately Uexküll never really offers sufficient reason to substantiate this claim. It is unclear, for instance, why Newton's science must imply the departure of God. It was a common argument in the eighteenth century to maintain that a perfectly causal world must have been created by God, rather than necessitate his absence. The argument was so well known that David Hume chose to embody it in the character of Cleanthes in his *Dialogues Concerning Natural Religion*. Cleanthes offers the hypothesis that the world is analogous to a great machine precisely to prove God's existence. Most of the dialogue is an entertaining and insightful confrontation between Cleanthes and Philo about this very proof and its ability and inability to explain both God and the world. Uexküll, it seems safe to say, believes that a causal, mechanical world only proves God's unimportance, not his existence. Nevertheless, it is his belief that with the rise of scientific reason in the eighteenth century there was a proportionate decline in finding meaning in a designed universe. One could still argue that God may have created the world and set it in motion, but with the increasingly pervasive belief in a perfectly rational and causal system, God was no longer necessary to keep it going. Like a machine, the earth and universe function perfectly well on their own without the creator. Thus, Uexküll concludes, "[t]he design of the world had broken down. Looking for it had become meaningless" (NCU, 114).

It is with the departure of God, then, that meaning unravels. But just as important, it is because of God's absence that Uexküll discovers the move from a mechanical and functional universe to one that is random and without plan. His narrative thus moves as follows: from a harmoniously designed universe (Kepler), to a meaningless mechanical system (Newton), to an accidental, planless world (Darwin). What is perplexing about each shift—what Thomas Kuhn would call "scientific revolutions"[6] of paradigms—is that God's absence underlies each one. How can God's departure be responsible for the movement from a designed universe, to a mechanical one, and to a planless one? The answer is not altogether clear and unfortunately one that Uexküll does not even begin to address. The theme of God reveals a religious current that runs through Uexküll's writings on nature and life. But for the purpose of understanding his position on nature's conformity with plan, it is not entirely necessary that his reasoning be complete in this story. What *is* important is how he finds Darwinian science as pervaded by a potentially harmful planlessness. With no God to oversee an inherently meaningful design in the world, nature might not be teleological after all.

As modern science cracks open the mysteries of nature, living things are gradually stripped of an inherent purpose. As Uexküll writes, "This way it became possible that not only the inorganic world, but also the living things were declared products of accidental happenings.... Finally man himself became an accidental product with purely mechanical, aimlessly functioning physical processes" (NCU, 115).

What Darwinian evolution promotes, according to Uexküll, is a "causal chain" that unfolds from "random displacements" and accidental occurrences, rather than an overarching design at work within nature. Uexküll echoes these reservations in his earlier work, *Theoretical Biology*, where he accuses Darwin of propagating "hopeless confusion" (264) with his theory of evolution. One of the problems is the purported misuse of the term "evolution." Uexküll explains that evolution derives from the Latin term *evolutio*, meaning an "unrolling" or "unfolding" (263). What he finds confusing is that rather than asserting a theory that details fewer folds (evolution as an unfolding of folds), Darwin's theory seems to advocate greater complexity by introducing more and more folds into the process. A paradox is seen between the etymology of evolution and its actual application; according to Uexküll, evolution ought to be a theory of increasingly fewer folds but instead becomes one of even greater complexity under Darwin. While this may be a false problem that Uexküll introduces—for he unconvincingly interprets 'unfolding' as being synonymous with less complication—it demonstrates the degree of his dislike for Darwinian evolution.

This said, the issue at hand has less to do with Uexküll's critique of Darwinism than with his promoting a different direction for biological theory. Kalevi Kull writes that "despite his opposition to Darwinism, Uexküll was not anti-evolutionist" so much as he was a firm proponent of epigenesis (Uexküll, 5). It is perhaps fair to say that he directed his attention more to the issue of physiological development than to evolution itself, even if, and particularly because, his remarks against evolutionary theory never appear convincing. His principal objection to this point is that "evolution means that within the germ the finished animal already lies concealed, just as the folded bud contains the perfect flower, and in addition to growing, has merely to unfold and evolve in order to produce it" (TB, 264). Aside from being reductive and misattributing the theory of preformationism to evolution, his interpretations often seem to be wrong and could be the result of having misunderstood Darwin's ideas.[7]

The focus on epigenesis offers a more informative look at how Uexküll distances himself from Darwinism. Whether correctly or not, Uexküll believes that Darwinian evolution offers a constantly changing horizon in which accidents occur and random pairings coincide to produce strange and potentially monstrous offspring. The accidents of natural history, together with the notion that once paired, only the parental ancestors contribute to an organism's development, lead Uexküll to believe that, on the one hand, Darwinism is too haphazard in accounting for natural events and, on the other, too concerned with purely material interactions between specific ancestors. In other words, he finds Darwinism too complacent in attributing nature's growth to random, historical chance and too materialistic in claiming that only the inheritance of 'genes' lead to the future of the species. This reading reinforces the "hopeless confusion" that he perceived in Darwin's ideas: both a chaotic freedom and a materialist determinism, both chance coincidences of a long history and the particular determinism of parental 'genes.' The result of such an interpretation is something akin to a planless, chaotic physicalism:

> Since Darwin's day, we see not only the inorganic objects, but also the living things in the sensed-worlds of our fellow-men, fall to pieces. In the majority of sensed-worlds, animals and plants have become nothing but assemblages of atoms without plan. The same process has also seized on the human being in the sensed-worlds, where even the subject's own body is just an assemblage of matter, and all its manifestations have become reduced to physical atomic processes. (TB, 335)[8]

The repercussions for Uexküll's own theory is that nature has more of a regulative plan than Darwin suggests, and that more than just the material genes of the two parents contribute to the development of organisms. His confrontations with Darwinism point toward his notion that nature has a conformity with plan.

To repeat, Uexküll's conformity with plan attempts to steer a path between the mechanical laws of chemistry and physics and the apparently random variations in nature suggested by Darwinism. For Uexküll, nature is neither entirely causal, nor is it just random; it is neither simply physical, nor is it spiritual. Rather, nature accords with an overarching plan that has set parameters in which life forms can interact (thus not entirely random) as well as inclusive of agents and forces other than the parental genes as developmentally constitutive for the organism (thus not exclusively materialistic or organic). To be fair, Uexküll paints an overly simplistic and one-sided picture of physics, chemistry, and Darwinism, as distinct ideologies as extreme in their views as they are wrong for biological science. I have drawn this comparison in order to better illustrate what Uexküll is working against in the formation of his own theoretical biology.

If nature's conformity with plan has little in common with either physics and chemistry or Darwinism, then how are we to understand it? Might this parting of ways signal a return to Kant's influence on Uexküll? It might be tempting to turn to Kant, particularly to his *Critique of Judgment* and the later writings on history where he expresses a teleological theory of nature. But despite his occasional appeals to Kant's philosophy, Uexküll does not follow his teleology. In fact, he explicitly renounces a teleological force behind nature: "Instead of seeing in it merely a rule stretching across time and space, men have spoken of 'purpose' and 'purposefulness' in Nature. . . . It is advisable therefore to dismiss from biology, for all time, expressions such as 'purpose' and 'purposefulness' " (TB, 270). What Uexküll finds problematic in teleology is its deceptive tendency to anthropomorphize nature; that is, to see nature as guided toward ends that only we humans can objectively perceive. This may account for why Uexküll allows for a "rule" to stretch across time and space, but not one that considers purposive ends. To see a purpose is to presume insight into the full working of nature and thus to also perhaps see where it is heading. If this were the case, we would not only have insight into nature as a whole, which presumes the absolute standpoint of physics that he has already dismissed, but also the ability to interfere and control nature's future. This would further suppose that nature's rules may be altered or changed. In contrast, the rules of nature's plan appear to be unalterable: "This force of Nature we have called conformity with 'plan' because we are able to follow it with our apperception only when it combines the manifold details into one whole by means of rules. Higher rules, which unite things separated even by time, are in general called plans, without any reference to whether they depend on human purposes or not" (TB, 175–76). One can see that Uexküll, despite his reservations with teleology, nevertheless remains Kantian in his language.

With this point, we begin to move away from his critical appraisal of other positions toward the establishing of his own theoretical contributions. It has already been mentioned that he favors a 'horizontal' view of nature as opposed to a 'vertical' one. The idea that nature conforms to a plan acquires its greatest support from Uexküll's observations of rules that extend horizontally across time and space, rather than as lineages descending historically through time. While demonstrating a reluctance to embrace Newtonian physics and Darwinian evolution to explain biological phenomena, his own position becomes increasingly interesting in how he extends his observations across the horizon of nature. Nature becomes akin to a "web of life" that extends in all directions uniting both living and nonliving things into a cohesive design. Uexküll expresses this idea in the following manner:

> These mutual restrictions give us proof that we have before us a coarse-meshed tissue, which can be comprehended only from a standpoint higher than those afforded us by individual, com-

munity, or species. This all-embracing interweaving cannot be referred to any particular formative impetus. Here at last we see the action of life as such, working in conformity with plan. (TB, 258)

It is with this "all-embracing interweaving" view of nature that Uexküll makes his greatest impact in the fields of ecology and ethology. Nature conforms to a plan, a "super-mechanical principle" (TB, 350), that has no "formative impetus," but that extends across all things, both organic and inorganic. To better understand nature's plan, or at least derive a better indication of its design, we now turn to Uexküll's groundbreaking studies of animal *Umwelten*. With his *Umwelt* research, we return to the Kantian notion from which we began—namely, that "all reality is subjective appearance"—as well as to an elucidation of the web-like forms of life that constitute animal environments.

UMWELTFORSCHUNG

Uexküll is probably best known for the advances he made in the study of animal behavior. His innovation was to approach the environments of animals as not only a feature of ethology but as absolutely necessary to understanding animal life. The animal, together with its environment, are observed to form a whole system that Uexküll called an *Umwelt*, a term that he popularized as early as 1909 in his book *The Environment and Inner World of Animals*. His studies eventually led him to establish the field of *Umweltforschung*, the research and study of animal environments, as a way for biology to become a science more true to the animal *as* a subject with its own experiences.

How the *Umwelt* became important to Uexküll's studies can be traced once again back to Kant. The degree to which Uexküll leans on his interpretation of Kant demonstrates just how informative Kant's philosophy was to his biology, even if Uexküll does not always appeal to him or even fully elucidate the finer details of Kant's system.[9] Nevertheless, the idea that "all reality is subjective appearance" informs all of Uexküll's thought, and it reappears as central to his discussions of *Umwelten*. As one indication, he notes that "Kant had already shaken the complacent position of the universe by exposing it as being merely a human form of perception. From there on it was a short step to reinstall the *Umwelt* space of the individual human being in its proper position" (IU, 109). It is not difficult to see why the concept of *Umwelt* became so important once reality is acknowledged as subjective appearance. If it is agreed that the world is constituted through each individual subject, then it becomes necessary to ask how the world appears to each organism as a subjective appearance. What quickly becomes clear is that it is no longer easy to speak of "the world" as an objective fact, as a reality independent of our subjective experiences. In a remarkable passage, we are informed that things in the world have no existence independent of our individual perceptions:

> Objects, equipped with all the possible sensory characteristics, always remain products of the human subject; they are not things that have an existence independent of the subject. They become 'things' in front of us only when they have become covered by all the sensory envelopes that the island of the senses can give them. What they were before that, before they became covered, is something we will never find out. (TB, 107)

If this is so—namely, that objects do not exist independent of subjects who sense them—then not only the things in the world but the world as such becomes a concept in need of clarification. This is precisely what Uexküll intends when he introduces the concept of the *Umwelt*—to differentiate it from the objective world—and in its application to all animal subjects and not just humans alone. As we shall see, these distinctions between Umwelt and world, on the one hand, and human and animal, on the other, hold particular significance.

In order to better appreciate the lives of animals, the environments in which they live require illustration. But what is an environment if not the subjective appearance of the animal in question? Does the environment just bring us back to the animal? In a passage that shows a certain affinity with Heidegger's notion of being-in-the-world, Uexküll suggests that the animal and *Umwelt* are not two distinct beings, but a unitary structure that must be considered holistically: "all things within [the plan] must react on one another. So we may begin either by studying subjects, or by investigating their appearance-worlds. The one could not exist without the other" (TB, 71).[10] If it is the case that each organism in effect creates its own environment, then it is plausible that there are just as many environments as there are organisms. Uexküll concludes as much when, in reference to the question of whether the world can only be known through human cognition, he writes that "this fallacy is fed by a belief in the existence of a single world, into which all living creatures are pigeonholed" (SAM, 14). There no more exists a single world than there exists a single organism that inhabits it. He argues just the opposite. In contrast to the physicists' world, which he claims to be but "one real world," Uexküll proudly claims that "the biologist, on the other hand, maintains that there are as many worlds as there are subjects"

(TB, 70). To substantiate this claim, he frequently appeals to examples drawn from his empirical research, such as a seemingly "objective" description of a meadow or a tree, only to break down the landscape into a multitude of different *Umwelten* according to each individual organism. In one example, Uexküll notes how even something as simple as a single flower, can be a sign of adornment for a human, a pipe full of liquid for an insect, a path to cross for the ant, or a source of nourishment for a cow (IU, 108; TM, 29). From the case of a single flower, it is easy to see how a tree, coral reef, underground soil, or, larger still, a meadow, forest, or ocean may prove to be composed of a wide diversity of *Umwelten*, rather than just one real world. In the case of each organism, a new world comes into being, and, with each new world, one finds a further demonstration of one of Uexküll's favorite metaphors for the *Umwelt*: the soap bubble.

The image of a soap bubble surrounding every living being may well be one of the most endearing aspects of Uexküll's thought. This metaphor describes how the spherical *Umwelt* circles around and contains the limits of each specific organism's life, cutting the organism off in two respects: it provides a limit to the bounds of the organism's environment, but also acts as a layer that shields the organism from our observation. This motif appears consistently in his literature, and it is one that plays a central role in later interpretations of him, specifically by Heidegger and Merleau-Ponty. Appealing to a self-enclosed sphere is itself not new to philosophical discourses, and it could be the case that Uexküll may even be drawing from Leibniz's theory of monads when describing the spherical *Umwelt* as a soap bubble, as has been suggested in an early commentary.[11] More generally, the notion of a spherical *Umwelt* may simply derive from a tradition that likens the natural world to such things as atoms, planets, orbs, and the solar system. The *Umwelt* might be considered as akin to a microcosm in this respect. Nevertheless, Uexküll was fond of the soap bubble image:

> the space peculiar to each animal, wherever that animal may be, can be compared to a soap bubble which completely surrounds the creature at a greater or less distance. The extended soap bubble constitutes the limit of what is finite for the animal, and therewith the limit of its world; what lies behind that is hidden in infinity. (TB, 42)

Perhaps most decisive in this description is not so much that we are meant to think of the organism as being encased within something akin to a literal bubble (though he does suggest as much), but that each organism is limited as to what is accessible to it. The *Umwelt* forms a figurative perimeter around the organism, 'inside' of which certain things are significant and meaningful, and 'outside' of which other things are as good as nonexistent insofar as they are "hidden in infinity."

A good example of this, and one that is frequently cited in literature on Uexküll, is his description of the tick (*Ixodes rhitinis*). The life of the tick, and the female tick more specifically, provides a useful illustration of an organism's *Umwelt* because of its relative simplicity and the ease with which it can be variously interpreted. "Out of the vast world which surrounds the tick," Uexküll claims, "three stimuli shine forth from the dark like beacons, and serve as guides to lead her unerringly to her goal" (SAM, 12). Nearly everything in the external world that surrounds the tick has no significance to it. The moon, weather, birds, noises, leaves, shadows, and so forth do not matter to the tick. They may belong to the *Umwelt* of other organisms that live in the midst of the tick, but they do not carry any meaning for the tick itself. The external world (*Welt*) is as good as nonexistent, as are the general surroundings (*Umgebung*) of the organism. Both are theoretical references to contrast with the meaningful world of the *Umwelt*. What does matter to the tick, however, is the sensory perception of heat and sweat from a warm-blooded animal, on which the female tick feeds, lays its eggs, and dies.

Uexküll recounts how ticks will position themselves in a hanging position on the tip of a tree branch in the anticipation of a mammal passing beneath the branch (SAM, 6–13). After mating, the blind and deaf tick is first drawn upward by the photoreceptivity of her skin. While the tick hangs on a branch, very little affects it. The tick does not feed itself, shelter itself, or engage in any other activities. It simply waits.[12] And, remarkably, ticks have been noted to hang motionless for up to eighteen years at a time until a precise environmental cue eventually triggers it from its rest. This span of time encompasses nearly the entire life span of the tick, and it does so until the tick senses a specific odor emanating from the butyric acid (sweat) of a mammal. This sensation triggers a second response: the tick releases itself from the branch in order to fall onto the hair of the moving mammal. At this point, the tick's third response is to turn toward the source of the heat and bore itself into the mammal's skin. The taste of the blood matters little; experiments have shown that the liquid has to be the right temperature in order for the tick to drink. These three cues (what Deleuze will call "affects") constitute the *Umwelt* of the tick: (1) drawn by the sun, it climbs to the tip of a branch, (2) sensing the heat of the mammal, the tick drops onto it, and (3) finding a hairless spot, the tick feeds on the mammal's blood. Once the tick has bored itself in, it sucks the mammal's blood until the warm blood reaches the tick's stomach, at which time a biological response is activated, and the sperm cells that a male has already deposited and are waiting in the female are released to fertilize the awaiting eggs. This reproductive action will not occur if the foregoing sequence of events first takes place.

At this point, the tick has accomplished its plan, and dies soon after. To be sure, many, if not most, ticks do not make it through this full cycle, but this does not diminish the significance of the tick's *Umwelt*. Above all else, these few environmental signs interest Uexküll the most. These signs alone constitute the *Umwelt* of a tick, such that everything else does not factor as meaningful in any way; indeed, there *is* nothing else for the tick, even if there may be for another organism. It is on this point that we can see a parallel with other organisms. In the way that a tick can sense the precise odor of mammalian sweat, the same odor may have no significance for other living beings. This sign does not figure into my *Umwelt*; it has no significance for me. However, I may perceive and be affected by the same mammal in another way. Perhaps the mammal is a dog out for a walk in the woods. Just as the mammal belongs within the *Umwelt* of the tick, the mammal may equally belong to my own *Umwelt*, albeit with a different significance. And while the dog may not notice the tick, it may notice a squirrel to chase or a twig to play with. With this understanding, it becomes clear how it can be said that these signs form the "soap bubble" in which this tick lives, in effect limiting the significance available to it. As Uexküll notes, "[e]ach *Umwelt* forms a closed unit in itself, which is governed, in all its parts, by the meaning it has for the subject" (TM, 30). But this example further demonstrates how the *Umwelten* of different organisms may overlap with one another. The relations between things expand and mesh with one another in the intricate web of life.

Before further addressing the role of significance and meaning—which become more central in Uexküll's later writings—one last important theme must be mentioned in relation to the *Umwelt*. Along with the metaphor of the soap bubble, Uexküll also frequently employs a musical reference to describe the *Umwelt*. However, whereas the soap bubble captures an organism's *Umwelt* by circling it within a defined parameter, the musical analogy extends outward by demonstrating how each organism enters into relationship with particular aspects of its surroundings. The two are not mutually exclusive, but rather offer complementary perspectives on the *Umwelt*. On the one hand, the soap bubble emphasizes how Uexküll sees the *Umwelt* as finite and spherical by encircling the organism within certain limits, and, just as important, precluding us from ever penetrating into another organism's soap bubble to fully understand the significance of its *Umwelt*. On the other hand, Uexküll characterizes nature as a harmony composed of different melodic and symphonic parts (TB, 29), such that the emphasis in this analogy is placed not on the limitations that capture the organism within a confined sphere, but with how organisms express themselves outwardly in the form of interlacing and contrapuntal relationships.

To better understand the workings of nature, it is therefore a matter of composing "a theory of the music of life" (NCU, 120). The music of life is roughly composed of five interconnected parts or segments. Although Uexküll is never completely explicit or consistent in his use of terminology, I believe we can nevertheless interpret his musical terminology with the following biological equivalents:

1. *Chime and/or rhythm of cells:* The basic form of music, a simple bell chime or rhythm, is found at the level of cellular movements. Since cells can be "subjects" in their own right, they too are capable and even necessary in forming a part of nature's music. For example, Uexküll writes: "The ego-qualities of these living bells made of nerve cells communicate with each other by means of rhythms and melodies: It is these melodies and rhythms that are made to resound in the Umwelt" (TM, 48).

2. *Melody of organs:* A melody is slightly more intricate than a rhythm, and thus belongs to the functioning of organs. For example, Uexküll writes: "The chime of the single-cell stage, which consisted of a disorderly ringing of single-cell bells, suddenly rings according to a uniform melody" (TM, 51). The melody of organs is best demonstrated in relation with the next stage:

3. *Symphony of the organism:* The organism as a whole works as a symphonic production of the different organ-melodies and cellular-rhythms that make it up. By adding the different chimes, rhythms, and melodies together, you get the symphony of an individual organism. For example, Uexküll writes: "the subject is progressively differentiated from cell-quality, through the melody of an organ to the symphony of the organism" (TM, 51).

4. *Harmony of organisms:* Harmony begins with at least two different living organisms acting in relation with one another, but harmony can also extend to a collective whole, such as a colony, swarm, herd, or pack. For example, Uexküll often notes the contrapuntal duet that forms a harmony between two organisms: "We see here [in pairs] the first comprehensive musical laws of nature. All living beings have their origin in a duet" (NCU, 118). Or: "two living organisms enter a harmonious, meaningful relationship with each other" (TM, 52). And further: "The harmony of performances is most clearly visible in the colonies of ants and honeybees. Here we have completely independent individuals that

keep up the life of the colony through the harmony of the individual performances [with each other]" (NCU, 118).

5. *Composition of nature:* When all of the parts of nature come together, it may be said that nature itself forms a musical composition. Although Uexküll is slightly hesitant in naming a precise composition of nature, he is no less certain that nature does form one: "Nature offers us no theories, so the expression 'a theory of the composition of nature' may be misleading. By such a theory is only meant a generalization of the rules that we believe we have discovered in the study of the composition of nature" (TM, 52).

While it is true that there can be many parts of nature that do not 'make music' with one another, Uexküll is nevertheless clear that despite any discordance, "disorderly ringing" among cells, or disharmony between organisms and things, nature as a whole exhibits an overall harmonic composition.

This theory of the harmonic composition of nature brings us back to the earlier expression of nature's "conformity with plan." If we recall Uexküll's antagonism toward the physicists' mechanical view of natural laws and their belief in the existence of one real world, we can now see how his theory of nature's musical composition is a response to it as well as a more unified formulation of his belief in nature's conformity with plan. "Instead of laws of mechanics," Uexküll explains, "the laws are here closer to the laws of musical harmonics. Thus the system of the elements starts with a dyad, followed by a triad, etc" (NCU, 116). Later in this same essay, he concludes this point when he notes how "we find all properties of living creatures connected to units according to a plan, and these units are contrapuntally matched to the properties of other units" (122). The plan that nature abides by is a musical score. Yet, Uexküll never to my knowledge confirms what type of musical score this might be. After all, to say that nature's plan is similar to a musical composition can conjure up many images of nature: is it a Vivaldian plan, with plenty of baroque orchestration? Or is nature more comparable to Schönberg's minimalist twelve-tone pieces? Or the off-tempered plays of a John Coltrane score? I would be curious to know what Uexküll might think of the experimental and chaotic score by Sylvano Bussoti that Deleuze and Guattari represent on the first page of their chapter "Introduction: Rhizome" in *A Thousand Plateaus*. Could such chaos be found within the overall ordered design of nature? Presumably a universal depiction of nature will always accommodate slices of chaos, just as we are left with the possibility of infinite subjective *Umwelten*. More than likely, Uexküll would respond that nature's compositional plan includes all of these scores, and many more.

The underlying point is not to suggest that he has not actually scored nature's composition—since he admits that nature offers us no precise theory—but to note that his theory is a response, even if not fully developed, to the plan offered by physics. The biological world of animals and their environments consists of an artful play of interconnections, to the degree that one organism is necessary for understanding an other. The *Umwelten* of organisms are therefore not simply closed spheres, as if locking the organism within a self-concealed and isolated container. The animal is not an object or entity, but a symphony underscored by rhythms and melodies reaching outward for greater accompaniment. Individual *Umwelten* are necessarily enmeshed with one another through a variety of relationships that create a harmonious whole. How the organic symphonies relate to one another to compose a grand design in nature is the subject of this next section.

BIOSEMIOTICS

In the end, why does it matter that Uexküll speaks about animals in terms of their respective *Umwelten*? And what is the significance of the two analogies, that of the soap bubble and the musical composition? In this final section, Uexküll's theories on life become truly interesting insofar as his ideas converge to express a unified view that accounts for the significance of all intermeshing environments.

Discussion on this issue really begins with the harmonious coupling of at least two organisms. It is true that Uexküll finds the musical analogy and soap bubble metaphor extending all the way back to individual cells. By itself, a single subject, whether an amoeba or mammal, may form a symphony of its parts, with each cell and organ ringing in melody with others to create an independently functioning organism. However, the subject is never really alone. It is only when it interacts with other things within an environmental setting that an understanding of the living being begins to emerge. Thus, the emphasis that we find in Uexküll's analysis is that rather than focusing on individual entities, which would just highlight a single tone ringing into an empty universe, we must come to recognize how each living thing only begins to show itself as part of a pair or as a duet. For life to commence, we need to start with a relation: "We see here the first comprehensive musical law of nature. All living beings have their origin in a duet" (NCU, 118).

In *The Natural Alien*, Neil Evernden underscored that Uexküll was one of the first to present "a biology of subjects" (78), such as we have seen, but I would like to push this further and suggest that what we have is an intersubjective theory of nature. Though Uexküll does not characterize his theory in phenomenological terms, it is clear that an intersubjective model is at its center. It is just that he speaks of the interrelations in terms of "counterpoint," "duets," and "harmonies," as opposed to the Other's gaze, consciousness, empathy, and intentionality, as one would find in the case of Husserl or Sartre. To be sure, there is great disparity between these theoretical orientations, but Uexküll is clear to highlight that the identity of an animal can never be approached other than through its intersubjective relations. He explains this in the following manner: "The theory of composition of music can serve as a model; it starts from the fact that at least two tones are needed to make harmony. In composing a duet, the two parts that are to blend into harmony must be written note for note and point for point with each other. On this principle the theory of counterpoint in music is based" (TM, 52). This theory serves as the model for his understanding of nature and it is one that emphasizes the need for at least two tones to create a meaningful picture.

In order to inquire into the biological world, therefore, we cannot begin with just a single organism any more than we can begin with the *Umwelt* alone. When framed in this way, an organism is never just one. Instead, each organism has a context, an *Umwelt* in which it lives, and, in being so, the organism is always already more than itself. It is the notion of the animal as "subject," then, that is precisely at issue. To know the organism requires knowing its other(s). But to what degree is the other, as other, a part of the subject? Where, in other words, does the subject begin or end, and likewise the environment? In *Theoretical Biology*, for instance, Uexküll suggests that we can and probably should consider the organism as resembling a community of subjects just as much as we think about a community or city like a large organism. This suggestion is not far off from various theoretical positions in the sciences today. The importance of boundaries in the ontological distinction of living beings is not necessarily new, but it is no less remarkable in this instance. Henri Bergson, in his introduction to *Creative Evolution*, forces us to confront the question of where individuality begins and ends. He concludes in a way that Uexküll would surely have appreciated: "In vain we force the living into this or that one of our molds. All the molds crack" (x). The difficulty of this question need not remain at the level of cells, organs, and animals either. For instance, Alphonso Lingis has pushed our thinking of the body and nature in his philosophical writings on packs, herds, and swarms; James Lovelock has famously argued in his Gaia theory that the planet Earth is a living organism; and, in a different manner, the physicist Lee Smolin has argued that the universe itself is like a product of evolution, continually developing and changing.[13] An issue here is that the contrapuntal arrangement can increasingly grow, such that the previous subject presumably becomes sublated within the next stage, from cell to animal to environment to ecosystem to planet to cosmos and so on. It is in this context that I find Uexküll's contrapuntal 'duet' becoming susceptible to a Hegelian dialectic, where each living subject needs another for its completion.[14] This likewise broaches the topic of what is actually living: what is the true subject of life? Answers to these questions are not easily forthcoming, nor does Uexküll even look to posit them in the first place. But even while his own thought remains at the level of counterpoints between animal and *Umwelt*, he has opened the question of the subject by situating it between animal and environment. He thus breaks the old mold and recasts it in a new light. Heidegger, Merleau-Ponty, and Deleuze await.

It is in this manner that meaning and significance surface as important concepts in Uexküll's later investigations. One of his final texts, *The Theory of Meaning*, outlines this importance most emphatically. It opens with the claim "that life can only be understood when one has acknowledged the importance of meaning [*Bedeutung*]" (26). He further claims that "*the question of meaning is, therefore, the crucial one to all living beings*" (37), as if to underscore that the question of meaning does not solely involve human relations, but that all living beings are said to generate their own unique meaning in terms of their respective *Umwelten*. With this in mind, we discover what Uexküll is truly after in his biological theories, including his concept of the *Umwelt* as a soap bubble and nature's overall conformity to a musical plan: the meaning of biology as a "theory of life" is to discover how meaning is generated through relationships. One may even be tempted to say that, in order to know a living being, one must know the relations it is capable of forming; an animal is no more than its relations. As we will see later, Deleuze claims as much in reference to Uexküll when he writes "an animal, a thing, is never separable from its relations with the world" (SPP, 168/125). Uexküll does not go so far as to suggest that organisms become fragmented subjects as a result of these relations; he still believes that there are essential "natures" to each living thing (TM, 72), even if the 'center' has been repositioned. However, the emphasis on outward relations plays a crucial role in accentuating how an organism is always more than itself by virtue of its symbiotic reciprocation in other things. These relations are thus the source for gaining access to the meaning of a given organism's life: "relations of meaning are the only true signposts in our exploration of *Umwelten*" (SAM, 40).

By placing an emphasis on meaning derived through biological signs, Uexküll's thought has come to be read as one of the foundations of biosemiotics (or zoosemiotics), the field of study that looks at how signs are communicated throughout living systems.[15] While biosemiotics has emerged as a discipline that studies such diverse phenomena as animal language and molecular genetics, one of its guiding principles is that living things are not purely mechanical processes, but messages to be read and interpreted. Thus it is that animal language can be studied as a sign system containing meaningful messages transmitted between organisms; likewise, DNA can be read, and is read, as message-bearing signs transmitted (via RNA) to protein for replication to occur.[16] Perhaps because of its interdisciplinary nature, the domain of biosemiotics has had a fairly marginal history and usually finds its home more often in semiotic studies than biology. Its relative neglect in biology is probably the result of the dualistic mentality within theoretical biology that pits two schools of thought against one another: a Baerian neovitalism that finds holistic processes within living systems and a Darwinian reductionism that considers evolution but forgets the organisms. However, the two schools are not entirely in opposition, and biosemiotics has slowly developed as a reputable field that studies, in a very broad manner, "the phenomena of recognition, memory, categorization, mimicry, learning, communication . . . together with the analysis of the application of the tools of semiotics (text, translation, interpretation, semiosis, types of signs, meaning) in the biological realm."[17]

Uexküll never used the term "biosemiotics" himself—it wasn't coined until 1961—and it is known that he was not familiar with the work of classical semioticians such as Peirce or Saussure.[18] From a biographical standpoint, it is particularly noteworthy that Uexküll was an acquaintance of the neo-Kantian philosopher Ernst Cassirer. Cassirer and Uexküll met in Hamburg while both had tenures at its university in the 1920s. While it is better known how Uexküll's biology found its way into Cassirer's thought, it is not as well known how much influence Cassirer may have had upon Uexküll.[19] Considering the influence and reputation of Cassirer during this time, it is highly probable that Uexküll was affected by Cassirer's neo-Kantianism as well as Cassirer's three-volume work on the *Philosophy of Symbolic Forms* (1923–1929). Insofar as both Kant and semiotics are central to Uexküll's theoretical biology, a case could be made for Cassirer's influence on the 'biosemiotic' thread in Uexküll's writings.

Nevertheless, despite his unfamiliarity with semiotics as such, Uexküll has been adopted retrospectively as one of the founders of biosemiotics by both semiotically inclined biologists (e.g., Salthe, Hoffmeyer, Lewontin, Kull) and biologically oriented semioticians (e.g., Sebeok, Deely). One of the main reasons for Uexküll's association with biosemiotics is the increasing emphasis he gave to nature as a system of signs. His studies of animal *Umwelten* gradually revealed what appeared to be a living play of signs and interpretations. As we have already seen, Uexküll was not particularly enamored with the mechanistic view of organisms. This position is further accentuated in his observations of how organisms act in a manner that can be attributed to neither instinctual, mechanical responses nor random acts.

Organisms, according to Uexküll, actively interpret their surroundings as replete with meaningful signs. They are not merely passive instruments or message bearers, but actively engaged in the creation of a significant environment. While it may be tempting to suggest that this is merely instinctual, this reductionist account does not take seriously the interpretive act on the part of the organism. Jesper Hoffmeyer explains that Uexküll's "whole point was that neither individual cells nor the organisms are passive pawns in the hands of external force [or, I would add, internal forces]. They create their own *Umwelt* and in so doing become a subjective part of Nature's grand design" (56). The creation of the *Umwelt* occurs through the interpretative work of the organism, whereby the organism responds to certain signs that are significant to it, and likewise creates signs for others. It is this focus on signs as meaningful within the construct of the *Umwelt* that has led to his association with biosemiotics.[20]

How does Uexküll's thought demonstrate biosemiotic ideas? Again, at the simplest level, it comes down to the pairing of at least two things. Life, it would seem, begins with two.[21] Within nature's grand design, Uexküll argues that organisms are responsive to certain signs that complement their own selves; the reflexivity of the sign with the organism leads to the collaborative production of the *Umwelt*. He has already explained that there are as many *Umwelten* as there are subjects, but now it is a matter of discerning how different *Umwelten* can resonate together in a meaningful and harmonious fashion. In this way, meaning thus becomes another way of referring to nature's harmony, albeit in more theoretical terms. "Meaning in nature's score," for example, "serves as a connecting link, or rather as a bridge, and takes the place of harmony in a musical score; it joins two of nature's factors" (TM, 64). What one looks for then are the relations that bridge and connect one thing with another. The relation or bridge is the generation of meaning itself.

We have already seen an example of this in the case of the tick and the mammal. Uexküll now introduces the technical terms of "meaning-receiver" for the tick and "meaning-carrier" to refer to the mammal, but we need not concern ourselves with the specific terminology here. Of primary interest is how the mammal serves as a "sign" (*das Zeichen*) for the tick to interpret and how this denotes a level of significance. The mammal serves as a significant counterpoint for the tick in that it elicits certain signs (odor of butyric acid, heat of blood) that become meaningful within the tick's *Umwelt*. The mammal emits a tone that complements the tick's own; a meaningful relation is formed. It is furthermore argued that the signs do not simply serve as externally causal forces, as though they were merely part of a mechanical order in the service of eliciting instinctual effects, but that the tick must actively 'interpret' the signs as being significant to it. Unfortunately,

Uexküll does not elaborate a theory of interpretation as such; the interpretive process remains a biological relation that occurs between an organism and its other, where neither is reducible to a cause–effect scenario. They both give and receive the sign of the other, and it is in the convergence of these signs that an interpretive process takes place. This example may be better illustrated by looking at another one.

Another example that Uexküll makes frequent use of is the flower as it appears to different organisms. In this scenario, he introduces a truly innovative way of looking at the relationship that develops between two things. To do so, he begins by reciting from a poem by Goethe—"If the eye were not sun-like, It could never behold the sun"—but then adds his own completion to this passage: "If the sun were not eye-like, It could not shine in any sky" (TM, 65). Uexküll draws inspiration from Goethe's insight and offers the following conclusion with respect to the flower and a bee:

> If the flower were not bee-like
> And the bee were not flower-like
> The unison could never be successful. (TM, 65)

The bee and the flower find a complement in each other insofar as one cannot be what it is without the other: the flower must be bee-like, and the bee must be flower-like. The two form a duet and together create a symbiotic relationship, where both the bee and the flower depend on one another for the maintenance of their individual livelihoods as well as that of their species; the bee needs the flower to collect nectar for the hive, and the flower needs the bee to help scatter its pollen. Lynn Margulis, a prominent molecular biologist, explains that such a union is a demonstration of "symbiosis," where one finds "the living together of very different kinds of organisms."[22] Such relations may even force us to reconsider how we classify and organize the natural world, an issue that Deleuze and Guattari promote in their readings of Uexküll. For the moment, however, it is at least worth noting that this complementary union between the bee and flower forms a symbiotic mesh linking the two together in a manner necessary for the survival of both. In a manner of speaking, the bee could not be what it is without the flower, and vice versa. Each depends on the other not only for survival but for the very way that we understand their lives.

Thus, there is something more at stake here than positing a symbiotic relationship. What Uexküll describes is a way of accentuating the relations between things, where the relations demonstrate a certain 'otherness' within each organism. The bee *is* flower-like and the flower *is* bee-like. If they were not, there would be little room for connection; each would pass by the other without significance. The only way that the bee and flower can have significance for the other is if each already *is* the other. Uexküll does not speak in ontological terms, but the descriptions that he offers have an ontological tone to them. The bee in a sense *becomes* flower-like and the flower *becomes* bee-like through the relation that they create together, but they can only become the other insofar as they already have an affinity for the other. They thus become a new ontological unit together, a meaningful system greater than their 'individual' parts. At bottom, an organism *is* what it is capable of becoming, insofar as it already is the other that it becomes in the harmonious relation.

Let us look at one more example, that of the spider and fly. Uexküll offers a similar account to frame how the spider has a "fly-likeness" that allows it to construct its spider's web according to the fly:

> The spider's web is certainly formed in a 'fly-like' manner, because the spider itself is 'fly-like.' To be 'fly-like' means that the body structure of the spider has taken on certain of the fly's characteristics—not from a specific fly, but rather from the fly's archetype. To express it more accurately, the spider's 'fly-likeness' comes about when its body structure has adopted certain themes from the fly's melody. (TM, 66)

This, again, is a striking depiction. The spider embodies "fly-likeness" within its structure. In a manner of speaking, the spider even anticipates the fly's presence through the fly's melody. It is well known how the spider spins its web in a dimension that perfectly eludes the perceptual dynamics of the fly's eye; the insect flies right into the web because it simply can't see the web. We can say, then, that the spider spins its web with a view toward the fly's coming presence; the web already captures fly-likeness before the fly even comes into the picture. Uexküll suggests as much by likening the spider's web to a painted work of art: "The web is truly a refined work of art that the spider has painted of the fly" (TM, 42). The mixing of metaphors in this aesthetics of life does not detract from the power of these descriptions. The spider embodies the fly, is fly-like, not because of some instinctive response, but because it has "adopted certain themes from the fly's melody." The spider has adapted itself to a meaningful sign in its *Umwelt* and has consequently become fly-like. The spider finds a counterpoint in the fly's melody, and strikes up a harmonious relation with the fly within its bodily structure and the spinning of the web. The web, in this instance, also provides a fitting metaphor for Uexküll's notion of the "web of life" as well as the meaning underlying the "biosemiotic web": the painted web that stretches to capture a contrapuntal melody brings an adhesive connectivity to this small domain of life and is representative of the interrelatedness of life itself.

Now, how can we better understand this sense of anticipation that an animal is said to embody? How does the bodily structure foretell the appearance of a significant other? Some have gone so far as to suggest that organisms exhibit a form of "intentionality," a concept usually found in theories of consciousness and the mind. There are many different definitions and usages of "intentionality" across the philosophical spectrum, from medieval scholasticism to Brentano to Dennett, but most share a similar understanding that intentionality implies being directed toward or about something. In phenomenology, for instance, Husserl noted that consciousness is always consciousness *of* something, that consciousness always has some object, state, feeling, and so on, that it is about. Since entertaining the difficult idea of animal minds and consciousness is not our present concern, how might Uexküll's thought prepare us to think of animals as intentional? Rather than appealing to consciousness, we might be better off considering intentionality by way of the moving and acting body. Jesper Hoffmeyer, for one, argues that Uexküll's biosemiotic analyses depict organisms as message bearers, which he takes to be an instance of *"evolutionary intentionality"* (47). Organisms, he writes, demonstrate an embodied "anticipatory power" for the "aboutness" of their own body and the environment. On this point, Hoffmeyer writes: "To say that living creatures harbor intentions is tantamount to saying that they can differentiate between phenomena in their surroundings and react to them selectively, as though some were better than others. Even an amoeba is capable of choosing to move in one direction rather than another" (47–48). While this usage of intentionality seems still too vague and preliminary, Hoffmeyer nevertheless pushes our interpretation of Uexküll to incorporate a bodily understanding of otherness. Even amoeba, he wants to say, anticipate their surroundings by interpreting cues and signs as meaningful, and thus they suggest a kind of intentionality toward their *Umwelt*, no matter how innocent and rudimentary this may be.

Meaning is acquired insofar as something shows itself within the *Umwelt*, where an otherwise "neutral object" becomes significant to the organism because it complements its own musical tone (TM, 27). This capacity is demonstrated by the degree to which the organism is capable of creating harmony with another tone. This capacity, moreover, is a part of the organism's very being, such that we may even speak of a spider's being 'fly-like' and its ability to become 'fly-like.' These are not traits that are acquired by the organism, but are anticipated and elicited through the sign of another being. Thus, one way to think of Uexküll's "anticipation" might be with Aristotle's concept of "potentiality": but an important difference here is that the spider has the potential to be 'fly-like,' not just 'spider-like.' The essentialism that Uexküll is susceptible to actually transcends species distinctions. To be a spider requires that it in fact become fly-like, for otherwise it is somehow not fulfilling its potential. Likewise, the tick must become mammalian, and the bee flower-like. The organism has the capacity to be other, albeit an other that fulfills its own melody.

Not only do signs transmit across living things within the environment, but the nonliving and inorganic play just as decisive a role. While this point may seem obvious, particularly insofar as we need only turn to our own lives to see how material and physical things form a part of our own selves, the consequences of what constitutes the inorganic is not always appreciated. Uexküll writes that "the properties of lifeless things also intervene contrapuntally in the design of living things" (NCU, 122), as well as that "life's conformity with plan embraces both inorganic and organic forces" (TB, 354). What I specifically want to draw attention to is that Uexküll not only emphasizes the role of lifeless *things*—including any physical and material thing such as a spider's web or any artificial product—but, just as important, inorganic *forces*—such as affects, temperatures, shadows, or noises. The forces that derive from inorganic things play a constitutive role in the formation of an organism and its *Umwelt*, and are just as essential to the organism as are material, physical things, whether they are organic or not.

We have seen this in the case of the tick, when the tick perceives the precise odor of a mammal's butyric acid. However, to be even more precise, it is a specific olfactory organ of the tick that perceives the sweat. Thus, the relation is not necessarily between the tick and the mammal (which we too easily assume), but between an organ and an odor. The tick doesn't perceive the mammal as a whole organism, let alone the mammal *as* a mammal; it doesn't even care what mammal it is, so long as it detects the precise odor that harmonizes with its sense organ. If this is the case—and Uexküll argues on the basis of scientific evidence that it is—then this has important ramifications for how we understand the ontology of relations and what the relations connect. What are at the ends or nodes of these relations? We assume things, entities, beings, substance. But it would seem that the relations do not involve "individuals" per se; instead they are a means of connecting an olfactory organ with a temperature, or a web with a line of flight, one melody and rhythm with another. By emphasizing relations, and the ontology of these relations, Uexküll opens the way for a critique of bodies as individual entities by way of the *Umwelt*.

CONCLUDING REMARKS

While delivering the Harvey Lectures at Harvard in 1958, Konrad Lorenz noted that the young science of ethology owed more to Uexküll than to any other person or school of behavior studies. Even though Lorenz had his differences with Uexküll's teachings, he never ceased to pay his debt—and ethology's debt—to the man whom he regarded as a pioneer in studying animal behavior. Coming from Lorenz, this is quite the acknowledgment, for it is Lorenz's own name that is synonymous with the science of ethology. This deferral of acknowledgment is all the more impressive when we consider what Lorenz applauds: the manner and execution of Uexküll's scientifically objective research; his insight and consideration of the animal as a subjective being in its own right; that there is no reality outside of the particular subjective environments of living beings; and that the animal and environment together consitute a whole unit and must be studied as such. From these reflections, I believe that we witness the rise of ethology along with, to use Karl Popper's phrase, a "biological ontology." Uexküll's particular perspective into the natural world comes to bear on how we think of the reality of the world as one grounded in the relation between living being and *Umwelt*. We have here the beginnings of an onto-ethology: an ontological elucidation of "what is" via the active behavior of living beings. Our glimpse into what it means to be an animal is arrived at through these relations.

This is not to say that Uexküll formulated his position in this way, or that he remained satisfied with the outcomes of his inquiries. As noted at the beginning, Uexküll considered his biology as departing in two ways from a Kantian foundation: one was toward the role played by the body in the understanding of the *Umwelt*, while the other was to look at the relations between the living being and the environment it inhabits. His adherence to Kantian philosophy is interesting, but in the end highlights one of the lingering problems in his biology. Although he took significant strides down each of these paths, it is also the case that the body and the nature of its relations continued to intrigue and perplex Uexküll throughout his career. He had hit upon a problem, but was unable to fully resolve it. What becomes of the animal body through its various relations? The meaning that an *Umwelt* holds will vary depending on the living being in question. But to what degree? Is there an ontological difference between beings? Despite having formulated that all reality is subjective appearance, the degree to which the body participates in the creation of these relations has not been fully considered. As he notes, "many problems await conceptual formulation, while others have not yet developed beyond the stage of formulating questions. Thus we know nothing so far of the extent to which the subject's own body enters into his *Umwelt*" (SAM, 73). We have already noted that the animal in question actually *becomes other* by virtue of the relations it is capable of forming. Just how much Uexküll opens the possibility for an ontological ethology will be explored throughout the ensuing chapters.

The relations are what is key in the reading that I will carry out in the following chapters. Heidegger is drawn to Uexküll's research because he finds in him an accomplice in biology in order to think through the concept of the world. But whereas Heidegger emphasizes the particular comportment of human Dasein as being-in-the-world, the behavior specific to animals is less clear. Animals, like human Dasein, have relations to their environments, but just how much do these relations indicate insight into the meaning of being? The metaphor of the soap bubble is particularly noteworthy as Heidegger narrates how animals are encircled within their spherical *Umwelten*; in contrast to the opening of the world instantiated through human existence, animals are declared poor in world inasmuch as other beings are never fully manifest to them. Animals remain captive to their surroundings, whereas humans can experience a releasement to the clearing of being. On this note, Heidegger will consider the conceptual differences between environment and world, between behavior and comportment, between living and existing, and between animal and human.

With Merleau-Ponty, the focus will centre more explicitly on the role of the body. Here the issue concerns how the animal body meshes with its world. Throughout his career, Merleau-Ponty suggests an engagement with Uexküll's biology because he sees the structure of behavior as demonstrating the phenomenon of "being-for-the-animal." Like Heidegger, Merleau-Ponty is also drawn to the soap bubble imagery, but Merleau-Ponty emphasizes even more so the musical dimension of existence. In his writings on nature, animal-being is characterized in terms of an underlying melodic and rhythmic connection to the environment. Merleau-Ponty inaugurates a new ontology to shed light on how bodies are not things but relations in the flesh of the world. Uexküll's writings play a role in this development.

In the case of Deleuze and Guattari, the main issue similarly falls on the nature of relations, but in such a way that they question the very meaning of the concept "body" and "organism." They find life to be a play of differential relations that form brief assemblages, where animal life is no longer akin to a sphere but punctuated lines of flight. The lines and planes that Deleuze and Guattari emphasize pose a risk to the spheres and circles that have so far dominated the picture linking the organism with environment. Uexküll becomes a Spinozist in their hands, as they seek to count the affective relations between different bodies. But bodies are unconventionally conceived as pertaining to anything capable of forming a relation. From Uexküll, we are introduced to an entirely new way of considering such things as milieus and territories, rhythms and refrains, and how becoming-animal questions previous ontological positions.

CHAPTER 2

Marking a Path into the Environments of Animals

Early in Martin Heidegger's landmark *Being and Time*, we find the following passage that is as innocuous as it is provocative:

> To talk about 'having an environment [*Umwelt*]' is ontically trivial, but ontologically it presents a problem. To solve it requires nothing else than defining the being of Dasein, and doing so in a way which is ontologically adequate. Although this state of being is one of which use has made in biology, especially since K. von Baer, one must not conclude that its philosophical use implies 'biologism.' For the environment is a structure which even biology as a positive science can never find and can never define, but must presuppose and constantly employ. (GA2, 58/84)

Clearly, Heidegger shows a concern with how this notion of 'having an environment' is being heedlessly misused, not the least of which among biologists. No one is specifically called out; it is only noted that, ever since Baer, this negligence has slowly become more audible. People are talking about our 'having an environment,' and they are doing so in a reckless manner. There are at least three components to this problem, and all of them will have to be answered in one way or another: (1) what does it mean *to have* an environment?; (2) what is an environment? is it different from the world?; and (3) to whom does 'having an environment' apply? Despite the claim that biology is unable to resolve this problem, and that no less than an analysis of human Dasein will do the trick, Heidegger never really manages to fully resolve the ontological problem of the environment in *Being and Time*. He will need to return to it and in a manner that deals with the problem head-on. We have not heard the last of the chatter surrounding biology's environments.

Before settling into Heidegger's writings on animal biology, it will be helpful to first situate these reflections within the context of his more influential writings on ontology. It is fairly well known that in his early attempt to elucidate fundamental ontology as the specific existential analytic of Dasein in *Being and Time*, Heidegger had to distance himself from more common understandings of ontology and human existence. While there is a long history of thinkers who have toiled over *what* something is (being as substance, entity, thing, object), at no point, Heidegger claims, has there been sufficient attention to *how* something *is* in such a way that it can *be* the being that it is. The history of beings had "forgotten" the more fundamental question of being.

Thus, in order to establish his account of human existence as distinct from the history of ontology, including from contemporary thinkers such as Dilthey, Husserl, Scheler, and Cassirer, Heidegger had to be clear about what the question of being entailed and, just as important, what it excluded. To be fair, the entirety of *Being and Time* is concerned with this project of clarifying the question of the meaning of being, so it is not my concern to address all the nuances of this great text. Instead, what is noteworthy for our present consideration is how Heidegger decisively cut off further investigation into the anthropological, psychological, or biological sides of human existence. If the question of being was to be addressed as such, particularly through the existential analytic of Dasein, this meant that his discourse needed to be cleared of any anthropological trace. Heidegger is not concerned with what a person is, whether it is a question of the person's biology, sexuality, ethnicity, class, consciousness, or some other substantial concern. These are all categories with a great deal of import, but they all treat the human as an ontic being; in other words, the human becomes a thing or object that can be studied irrespective of the manner by which it *is* the being that it is. Heidegger notes that this anthropological bias is the result of two dominant traditions of Western thought, Greek philosophy and Christianity:

> The two sources which are relevant for the traditional anthropology—the Greek definition and the clue which theology has provided—indicate that over and above the attempt to determine the essence of 'man' as an entity, the question of his being has remained forgotten, and that this being is rather conceived as something obvious or 'self-evident' in the sense of the *Being-present-at-hand* of other created Things. (GA2, 49/75)

The issue here is that previous accounts, whether anthropological, psychological, or biological, ignore the more fundamental question of how humans exist so that they may even be taken as beings among other beings.

Again, this distinction that Heidegger makes, namely the ontological difference between being and beings, is a significantly large one, to say the least. The access to the question of being, Heidegger explains, can only be found in "the *existential analytic of Dasein*," "from which all other ontologies can take their rise" (GA2, 13/34). For this reason, Heidegger, still early in chapter one of *Being and Time* (§10), separates his existential analysis of human Dasein from previous analyses of human existence that perceive human beings as a reified entity subject to study. Why is it important to note this here? There are a number of reasons, but primarily because he dismisses biology as an inappropriate domain for questioning the ontological foundations of human existence. Biology has very little, if anything, to offer fundamental ontology. This is not so much a judgment of value—Heidegger does not disparage the findings of these sciences per se—as it is an issue of priority. As Karl Löwith, a former student of Heidegger, puts it, fundamental ontology simply takes precedence over every other question: "While essence refers to the conceivable *what* I am, existence refers to the factual *that* I am and have-to-be. This *that* in man's existence precedes whatever he is, biologically, psychologically, socially."[1] It is the meaning of *being*, therefore, that requires attention before further inquiry into what a being is. But even if biology offers little, at least to begin with, to the clarification of why and how one exists, the same does not hold true of the opposite. The ontology of Dasein can provide a foundation for later inquiries into biology as a "science of life," as long as we are clear on what comes first: "The existential analytic of Dasein comes *before* any psychology or anthropology, and certainly before any biology" (GA2, 45/71). So while Heidegger remains true to his intentions by carrying out an existential analysis of Dasein in *Being and Time*, he also leaves this other venue open to future scrutiny so long as the existential analytic has already been identified, if not completed.

That Heidegger does not entirely close off the life sciences is important, for it gives him the opportunity to later pursue a comparative examination between Dasein and the domain of animal life, such as he does in his 1929–1930 lecture course, *The Fundamental Concepts of Metaphysics*. But before looking more closely at this lecture course, one further note needs to be made concerning Heidegger's distinction between his existential analytic and the life sciences. The issue boils down to two central differences. The first is that which I have already briefly addressed and will continue to address: that Dasein cannot be analyzed as if it were a present-at-hand 'thing.' Human Dasein is not a substance or object, but a way of being. To arrive at the meaning of being ("what is being?") requires a lengthy analysis of the being (Dasein) who can ask this question in the first place. The second issue surrounds another, though not unrelated, difference: Heidegger's conceptual distinction between Dasein's *existence* and the *lives* of other living beings. Dasein, we are informed, is not merely alive, but exists in a manner irreducible to the living:

> Life, in its own right, is a kind of being [*Seinsart*]; but essentially it is accessible only in Dasein. The ontology of life is accomplished by way of a privative Interpretation; it determines what must be the case if there can be anything like mere-aliveness. Life is not a mere Being-present-at-hand, nor is it Dasein. In turn, Dasein is never defined ontologically by regarding it as life (in an ontologically indefinite manner) plus something else. (GA2, 50/75)

This is an important passage that has deservedly received plenty of commentary.[2] Since I will later need to address the essential difference that Heidegger draws between life and existence, I will simply point out that, in this section of *Being and Time*, it is not only the field of ontic beings that are excluded from his initial analyses, but that "life" is too.[3] An interpretation of Dasein cannot be achieved by merely working off the basis of life, since Dasein itself must be analyzed before any definition of life can ever be properly attained. This may appear to be begging the question, but, for the moment, let us note that it simply reiterates Heidegger's claim that an existential analytic of Dasein must be prepared before any further inquiries into other beings, whether they are living or not.

Now how does an analysis of human Dasein get us back to a philosophical biology of animal life? More pointedly, can and should it? So far it would appear not. But this was not entirely the case. Even though *Being and Time* explicitly excludes discussion of human existence along the traditional lines of life-philosophy, the issue of living beings is never far off. This may seem paradoxical given that the question of whether Dasein is even remotely associated with other living things is cut off before it can even be posed. Kant's famous question "What is man?," as found in his *Logic*, consumed philosophical thought over the next two centuries, but is not answered in any one formulaic way by Heidegger.[4] As we have seen, the question has even been undermined by another, so much so that human Dasein's relation to the rest of nature and life is under threat. Defining the human as a rational animal, political animal, hairless bipedal mammal, self-conscious, cultural being, or what have you, are all unsatisfactory and insufficient. If this is the case, then so too are questions of the following type: Is Dasein related—essentially, existentially, ontologically—with our biological and evolutionary cousins, the chimpanzees, gorillas, and orangutans? What about with other beings, such as horses, whales, bees, and multicellular or unicellular life forms? Or what about beings that are not commonly thought of as alive, such as rocks, cars, or computers? Of course, as we now know, these are secondary concerns to the more pressing question of *how* Dasein *is*, such that these other concerns can even come to be of concern to Dasein. If there is not a relation between Dasein and animals, plants, or rocks, then the question will be how and where this relation breaks. We will also have to consider if Heidegger's ontological inquiry necessarily diminishes the relevance of these questions surrounding the biological kinship between humans and animals.

I pose the distinctions in this manner in order to highlight all the better how remarkable it is that these secondary concerns eventually do receive attention, and only a few years after *Being and Time*. However, in his 1929–1930 lecture course, Heidegger submits these 'metontological' themes for questioning in order to better clarify Dasein and the concept of world.[5] It is impressive that not only does Heidegger begin the course by answering the question "What is man?" with the indecisive "We do not yet know" (GA29/30, 10/7), but that he also posits this question of what it means to be-in-the-world by means of a comparative examination with animals, plants, and material substances. Thus, what was once exclusively cut off as secondary at best, reemerges in this lecture course as playing a pivotal role.

To avoid confusion, Heidegger does not promote ontic questions to the status of fundamental ontology. Animals, plants, and rocks are not raised to the level of either the meaning of being or the existential analytic of Dasein. However, within the 1929–1930 course, Heidegger admits that they may very well provide a more accurate glimpse of Dasein through the consideration of the concept of world. Insofar as other beings are found in the midst of the world Dasein inhabits, they become worthy of speculation. In terms of our considerations here, Heidegger's foray into what approximates a philosophy of biology is highly significant. For it is here that Heidegger, the philosopher famous for his purportedly abstract analyses of being and time, on the essence of thought and metaphysics, ventures into the domain of other beings and their environments, in all their biological and material splendor. The question is no longer that of 'what is an animal?' but the ontological question of 'what does it mean *to be* animal?'. Of greatest interest will be his descriptions of animal life (where a variety of new Heideggerian concepts emerge), particularly in terms of his ontological trinity regarding beings. I refer here to his theses that "the stone (material object) is *worldless*; the animal is *poor in world*; man is *world-forming*" (GA29/30, 263/177). Together this triptych reads as a fairly comprehensive view of beings. Yet it is precisely how these beings relate to the world that is in question as Heidegger works to disclose the essence of their respective being. In order to get there, we will have to consider brief accounts of contemporary biologists (such as those of Darwin, Baer, Buytendijk, Driesch, and Uexküll), Heidegger's three paths into the meaning of the world, how animals are defined essentially, and how animals, in comparison to Dasein, might be said to 'have an environment,' that difficult though important ontological problem. The framework of this entire discussion is the relational character of life that, like the question of being itself, has often been overlooked.

THE ESSENTIAL APPROACH TO THE ANIMAL

Toward the end of the long and pivotal fourth chapter in *The Fundamental Concepts of Metaphysics*, Heidegger admits that the analysis he just submitted on the essence of life does not offer a "definitive clarification of the *essence of animality*" (378/260). On first reading, this admission seems like a humble one. Heidegger, one must think, is simply distancing himself from having laid out a lengthy account on the essence of animality, particularly since the animal is not really his main concern. The life of the animal was merely an important and helpful hurdle to cross on the way toward further clarifying Dasein's being-in-the-world. However, on closer reading, it is not the essence of animality that Heidegger is unsure of, but simply its clarification—that is, the manner in which the essence has been represented. Perhaps sensing that he has already spent enough time on the biological animal—a little later he notes that he will forego pursuing "the history of biology from its beginnings up to the present" even though "it would be instructive" (379/261)—Heidegger responds that "we do not mean to imply that this represents the definitive clarification of the *essence of animality*" (378/260). So it is not that Heidegger hasn't offered a definitive essence of animality, but that he may not have offered its full clarification. The essence seems right, though it may only be a characterization, no matter how concrete it may be.

In an unusual manner, Heidegger is almost deferential to the sciences when it comes to animal life. He is nearly always critical, but he is also mindful of how the sciences would rightfully think little of a philosophy that immodestly tries to replicate its discoveries. Rather, Heidegger's analyses of the essence of animal life is not a repudiation of biological investigations—he really isn't qualified to pass judgment on their results—but what he repeatedly calls a "transformation of seeing and questioning" in science. In pursuing the essence of animality, Heidegger wants us to believe that he is merely offering a new perspective on what the sciences already intrinsically know but have not yet brought to light. Even if it is a new perspective, do Heidegger's analyses of animal life really differ so dramatically from biological studies? In a word, yes, since he questions the ontological essence of animal being. For one, we are shown an alternative to "the prevailing mechanistic and physicalist approach to nature" (378/260) that dominated nineteenth and early twentieth century biology. To this, we can also add that Heidegger's analysis contrasts with vitalism as well. Around the same time as Heidegger's 1929–1930 course, researchers in biology were undergoing a period of crisis in searching for other alternatives to the increasingly stale debate between mechanism and vitalism. Many new terms and schools were being introduced to overcome this dichotomy. Donna Haraway, for example, has recounted how "organicism" became one such option.[6] Another one, of course, was found in Uexküll's work, to which we will turn shortly. The variety of theories mirrored the subject of study, the difficult-to-define organism, the basic unit of life. Is the organism like a machine? Is there an inner, immaterial force driving the organism? Are we perhaps not looking at the organism in the right way? No single option had satisfactorily or sufficiently proven its case for understanding the living organism.

To the mix of these responses, Heidegger includes his own original voice: "Originality consists in nothing other than decisively seeing and thinking once again at the right moment of vision [*Augenblick*] that which is essential, that which has already been repeatedly seen and thought before" (GA29/30, 378/260). To instigate a transformation of contemporary biology, Heidegger clearly emphasizes that it cannot simply be a case of finding new "facts" that will somehow change the shape of biology. Rather, it is the opposite that is true: by learning to observe, think about, and question biological phenomena in new ways, we will in turn be led to novel perspectives on the already existing facts. Indeed, with this transformation of seeing, the so-called facts themselves will surely change as well. Next to mechanism and vitalism, therefore, we find ourselves in need of philosophical analyses of the essence of the animal. Throughout the course of his lectures Heidegger discovers the essence of animality and the essence of Dasein to be in the nature of their different relations to world. It is precisely in how these relations to world are understood that marks Heidegger's departure from mechanism and vitalism, as well as how he comes to find an ally in Uexküll's biological studies. But before we look more closely at Heidegger's own approach, we need to first consider how he interprets some of the more dominant biological theories of his time. Even though his references are often brief, they are nevertheless helpful in better situating his own foray into the domain of animal being.

HEIDEGGER AND THE BIOLOGISTS

Even though the 1929–1930 course demonstrates Heidegger's most sustained reflection on the philosophy of biology, the biologists themselves do not receive much recognition. And even though what he does say may not be so illuminating in terms of their respective positions, his remarks are nevertheless indicative of his own stance toward the issues. For better or worse, here are some small portraits that one can garner from a reading of Heidegger on the writings of biologists. I have only included either those who are most often mentioned, and specifically in relation to the environmental world, or those who are most well known. I therefore do not include every biologist Heidegger happens to treat.

Karl Ernst von Baer (1792–1876)

Baer appears so infrequently in Heidegger's writings that he is almost not worth mentioning, but since we have already observed his importance in biology, including his influence on Uexküll, he rightfully deserves a place. In his brief references, Heidegger believes Baer was on the right track in his research before being too hastily "buried" by Darwinism, now consigned to that part of human history that was once buried but awaits being uncovered again. He writes: "It is true that one scientist of the grand style, Karl Ernst von Baer, was able to see something essential in the first half of the last [nineteenth] century, even though it remains concealed within modern philosophical and theological perspectives" (GA29/30, 378/260). Heidegger does not elaborate on what Baer specifically saw—only that he was able to see something essential. We might hazard a guess that Heidegger recognized Baer's emphasis on the epigenetic development of the organism that unfolds according to a specific building plan. Development was always considered as unfolding as a whole, not in terms of parts. In both *Being and Time* and his course on *Logic*, it is suggested that Baer was one of the first to talk about the structural relation between animals and environments, but, unlike Uexküll, this was never a thematic feature of his thought (GA2, 58/84; GA21, 215–16). What is interesting, either way you read it, is that in the context of Heidegger and his interest in the essence of the organism, he finds somewhat of a precursor in Baer's ability "to see something essential" too, even if Baer's views turn out to be buried for what may be good scientific reasons.

Charles Darwin (1809–1882)

Heidegger's position toward Darwin is more interesting, if only because he often appears to be at odds with Darwin's thought. It is probably the case that he is merely critical of Darwin's methodology more than he is dismissive of the scientific and theoretical results that it produces. But when Heidegger makes claims that what is *essentially* true of one animal holds true of all animals as a universal thesis (GA29/30, 275/186), this suggests a peculiar reading of evolutionary history. When taken according to their essence, all animals are the same. If this is the case, however, where does natural selection and evolution fit in? How can essential features change and alter over time and space? Daniel Dennett, for instance, argues in *Darwin's Dangerous Idea* that the two views are incompatible; essence and evolution do not mix well. While there is certainly something to Dennett's anti-essentialist critique, it ultimately does not speak to Heidegger's approach. Heidegger does not hold that each animal is defined in a manner akin to metaphysical substrates, but rather with respect to an ontological standpoint.

Next to Baer, Darwin appears the least frequently in Heidegger's text, primarily due to two problems that Heidegger finds. The first has to do with methodology. Darwin's analysis of organisms is found to be too mechanical in how it dismantles the organism without respect for keeping the organism as a whole in mind. In a sense, Darwin is accused of focusing too narrowly on the parts of the organism to the exclusion of the whole. "The movement of Darwinism and the increasingly powerful, analytical method in morphology and physiology," Heidegger writes, is guilty in believing "that we can build up the organism through recourse to its elementary constituents without first having grasped the building plan, i.e., the essence of the organism, in its fundamental structure and without keeping this structure in view as that which guides the construction" (378/260). In not focusing enough on the underlying essence, or in this case, on the organism as a whole, Heidegger finds Darwin's attention misguided. Darwin has broken the organism into elementary parts, and we are led to imagine that he has neglected the question of what it means to be an animal in the first place.

If the first problem is in Darwin's methodology, the second problem is in how Darwin conceives of the animal as a being. The two problems are not unrelated. Heidegger recognizes that Darwin acknowledges the environment as playing a role in his scientific theories of the animal, but it is in how the animal is related to the environment that is at issue. At bottom, Darwin is guilty of treating the animal as an entity "present at hand" within the environment insofar as he studies animals as entities and only secondarily does he consider their manner of being. Darwinism never fully considers the intrinsic relation between the animal and the environment, as Heidegger writes here: "In Darwinism such investigations were based upon the fundamentally misconceived idea that the animal is present at hand, and then subsequently adapts itself to a world that is present at hand, that it then comports itself accordingly and that the fittest individual gets selected"

(382/263). Darwin approached life too "materially": from a reductive view, he had an animal and an environment, added them together, and came up with a result. What is missing is precisely the nature of this relational structure between animal and environment such that the two may be understood as *essentially* related. It is not enough to just throw the two things together and hope to produce a result. Rather, Heidegger's contention is that in order to understand the animal and the environment, one must be able to account for the relation itself. This understanding of the relation he finds lacking in Darwin's account of natural selection.

This view is echoed in a brief overview of the developmental theory of adaptation. Without naming Darwin per se, Heidegger describes in a few sentences how animals most successful at adapting themselves to their conditions represent a "survival of the best" leading toward an "increasing perfection" of "higher animal species . . . out of primeval slime [*Urschleim*]" (402/277). It may be a caricature of Darwin's theory, but Heidegger's point is that this theory rests on an "impossible presupposition": "the presupposition that beings as such are given to all animals and moreover given to them all in the same intrinsic way, so that all the animal has to do is to adapt itself accordingly. But," Heidegger continues, "this view collapses once we understand animals and animal being from out of the essence of animality" (402/277). Heidegger's problem with this position is that it claims all animals have equal access to all other beings and adapt themselves accordingly. As we will see, however, one of Heidegger's principle tenets is that not everything discloses itself as such or to all animals equally. Animals may adapt in their struggle for survival, but they do not struggle against a world of other beings fully present to them as such. Some beings simply do not appear as significant to certain animals (they do not 'show' themselves within the environment, even though they are physically present), and those that *are* given are given in a manifold of different ways. Not all animals have access to all things in the same way. In Darwin's defense, he does not claim this anyhow; animals adapt within the confines of their own lives, and do so to unequal advantages. This would be an underlying point behind natural selection and the phenomenon of extinction: some organisms and species simply do not adapt as well as others because they do not relate to their surroundings in the "same intrinsic way."

Given Heidegger's stance toward Darwinism, it comes as all the more surprising that Emmanuel Levinas, a former student of Heidegger's, would liken Heidegger's thought with Darwin's. Like Heidegger, Levinas claims that human beings are a phenomenon distinct from the animal. Unlike Heidegger, however, who never directly argues against evolutionary theory (it never comes up as such), Levinas pits himself against the notion that humans are "only the last stage of the evolution of the animal." He continues:

I do not know at what moment the human appears, but what I want to emphasize is that the human breaks with pure being. A being is something that is attached to being, to its own being. That is Darwin's idea. The being of animals is a struggle for life. A struggle for life without ethics. It is a question of might. Heidegger says at the beginning of *Being and Time* that *Dasein* is a being who in his being is concerned for this being itself. That's Darwin's idea: the living being struggles for life. The aim of being is being itself.[7]

He breaks off here to get back to clarifying his own philosophy, but, in this brief description, which is admittedly only an interview, Levinas depicts a puzzling comparison between Heidegger and Darwin. Levinas's readings of Heidegger are usually much more careful than is shown here, for in this lapse he equates Dasein with a living being (i.e., an animal), and Dasein's *concern* for being (*Sein*) with an animal's *struggle* for survival. It is a perplexing passage, particularly when read in conjunction with Heidegger's own thought on adaptation and the struggle for survival. As we will see later, Heidegger would not disagree that animals struggle within their lives. However, this struggle is not the battle for life but an essential feature of being for the animal—a struggle against the dictates of the animal's own being (GA29/30, 374/257).

F. J. J. Buytendijk (1887–1974)

The Dutch biologist F. J. J. Butendijk presents a unique case in that his treatment by Heidegger provides a rare look at Heidegger's reflections on the body, a topic about which Heidegger is notoriously difficult. Some critics argue that he is far too abstract in his accounts of human Dasein. Hans Jonas, for instance, despite the influence of his former mentor, nevertheless challenges Heidegger on the lack of corporeality in his descriptions of human mortality. "Is the body ever mentioned?," he asks in his paper "Philosophy at the End of the Century."[8] Does it not suffer any physical needs? Almost in spite of the thrilling descriptions that Heidegger offers of human existence, Dasein appeared to be elevated above its corporeal and natural base. Parenthetical remarks like "Dasein's 'bodily nature' hides a whole problematic of its own, though we shall not treat it here" (GA2, 108/143) certainly don't help. Others, however, find Heidegger's thought to be full of bodily materiality; it may not be explicit, but it is always there. Frank Schalow has most recently argued to this end.[9] But it is here in the realm of animal biology that one expects to find some commentary on the body, and it is in Heidegger's reply to Buytendijk that we discover it. As we shall later see, Buytendijk was also influential in the development of Merleau-Ponty's thought, as particularly seen in *The Structure of Behavior*, where his phenomenology of the body is already evident.

In their brief encounter, Heidegger takes exception with Buytendijk's analysis of the animal's bodily contact with the environment. Heidegger quotes from *Investigations on the Essential Differences between Humans and Animals* wherein Buytendijk writes that the animal's body is bound to the environment almost as intimately as the unity of the body itself.[10] At issue for Buytendijk is how closely to think of this parallel between the cohesiveness of the body with the environment and the inner cohesiveness of the body's organs with one another. (One can see how Merleau-Ponty would be interested in this discussion, particularly in his formulation of "flesh" in *The Visible and the Invisible*.) Heidegger's problem with Buytendijk's formulation is that it seems to imply two separate relations that Buytendijk attempts to make analogous: body–environment and organ–organ within the body. For Heidegger, however, the two relations are not merely analogous; they are intertwined in the very behavior of the animal such that the 'two' relations are imbricated in a single understanding of an animal's being. Heidegger writes: "Against this [i.e., Buytendijk's position] we must say that the way in which the animal is bound to its environment is not merely almost as intimate, or even as intimate, as the unity of the body but rather that the unity of the animal's body is grounded as a unified animal body precisely in the *unity of captivation*" (GA29/30, 376/258). In other words, the body is a unified whole only because of its particular immersion in the environment, not vice versa. The body's relation to the environment reflexively establishes the body in its unity, and with the environment. Thus, there can only be a body if it is bound up with the environment; this relation is fundamental to any postulation of 'unity.' We have here the beginnings of a new conception of bodily being, even if it is still not fully developed. This is an important claim that comes after much of Heidegger's working through animal life, so let us leave a further examination until later.

Hans Driesch (1867–1941)

Along with Uexküll, Driesch is the sole biologist to be named in one of Heidegger's section headings as providing an essential step in biology. Although Heidegger claims that the results of Driesch's investigations are "no longer conclusive today," they still merit an exclusive place in his analyses. Driesch offers the first of two essential steps, namely, "the holistic character of the organism." Through his research with the embryos of sea urchins, Driesch's insight concerned how individual cells within the developing embryo unfold in relation to the organism as a whole. Ironically, this discovery proved to be both the benefit and the bane of Driesch's theory. What Heidegger finds to be of decisive significance is Driesch's focus on the organism as a whole, rather than thinking of the organism as an aggregate or composite of functional parts. "What is essential," Heidegger writes, "is simply the fact that the organism as such asserts itself at every stage of life of the living being. Its unity and wholeness is not the subsequent result of proven interconnections" (GA29/30, 382/262). Seeing the organism as a whole will prove to be not only essential to Heidegger's own analysis of organisms, it will also be what was so problematic with Driesch's attempt to escape mechanistic interpretations of living things. In seeing the organism as a whole rather than as a composite of parts, Driesch opened the door to the accusation of neovitalism, which Heidegger points out as "a great danger" of his theory. For in demonstrating the wholeness of the organism's embryological development, Driesch also advocated a teleological force that drives the development toward its purposive end. In forwarding this view, Driesch is charged with reinstituting "the old conception of life" that slips in a mysterious acting cause within the structure of the organism. Take the following, for example, as a representative statement from Driesch's 1907–1908 Gifford Lectures: "In this way, then, we finally get all phenomena in the living being which can be shown to be directed to a single point, thought of in some sense as an end, subordinated to the *purely descriptive* concept of purposiveness."[11] The organism is viewed as a whole at every stage of its development, but the whole is driven by some immaterial cause. A neovitalist claim of this sort, Heidegger remarks, is just as dangerous as the mechanist view it seeks to replace (GA29/30, 381/262).

Not only does Driesch's emphasis on the organic whole admit the dangers of a neovitalism, he also fails to appropriately conceive the organism in its environment. Driesch, in other words, while perceptively illustrating the whole of the organism at each and every stage of development, does not adequately relate the organic whole to the environment in which it is situated. "The animal's relation to the environment," Heidegger writes, "has not been included in the fundamental structure of the organism. The totality of the organism coincides as it were with the external surface of the animal's body" (382/262). In a manner similar to Buytendijk, the encompassing whole of the organism's life excludes a full interrelation with the environment. The organism is self-sufficiently whole irrespective of the environment in which it lives. The surface of the body—skin, hair, fur, scale, exoskeleton, feather, or otherwise—provides the outer limit and boundary of the organism. This oversight offers a further problem that Heidegger finds in his brief encounter with Driesch's biology: he has already indicated that he expects more from this relation. Thankfully, he finds a remedy in the second essential step provided by Uexküll's theory of animal life.

Jakob von Uexküll (1864–1944)

One could argue that Uexküll offers a more rounded account of the organism than do Driesch, Darwin, or Buytendijk. On the one hand, Driesch emphasizes the organism as a whole, rather than as an aggregate of parts, but he is also susceptible to holding the organism as a purposeful living being. He is congratulated for the former insight, but criticized just as much for not going far enough and for admitting neovitalism via the backdoor. Buytendijk, on the other hand, can be read as developing Driesch's theory insofar as he interprets the organism as a whole but in that he also offers an account of the environment's influence. Where Buytendijk's theory becomes problematic, however, is that in Heidegger's reading both the organism and the environment remain two separate entities that are only related in a subsequent and additive manner. This critique is similar to that accorded to Darwin. The organism and environment are added together in the hope of constructing a unique equation between the two. But the organism and environment cannot simply be appended to one another as a second thought. It is in adding an alternative to this scenario that Uexküll becomes important, not just for Heidegger's reading but within the history of biology.

Following Driesch's first step, Heidegger credits Uexküll with providing a second essential step in biology: his "insight into the *relational structure between the animal and its environment*" (GA29/30, 382/263). The nature of this relation is precisely what is at stake in these different theories of the organism. Uexküll's contribution lies in that, as we saw in the last chapter, he thematizes the relational structure as inherently necessary to understanding both organism and environment. Insofar as he does so, Uexküll renews the ecological dimension of biological studies through "the astonishing sureness and abundance of his observations and his appropriate descriptions" of how "animals are at home in the world" (383/263). In fact, Heidegger has very little to criticize in Uexküll's descriptions of animals and praises him as "one of the most perceptive of contemporary biologists" (315/215). More so than any other biologist mentioned within Heidegger's writings, Uexküll receives priority of place from the beginning to the end of these lectures on animal life. The following assertion is one example: "It would be foolish if we attempted to impute or ascribe philosophical inadequacy to Uexküll's interpretations, instead of recognizing that the engagement with concrete investigations like this is one of the most fruitful things that philosophy can learn from contemporary biology" (383/263). Heidegger was so taken by Uexküll's advances that he references him some ten years later in his 1939 graduate seminar on Herder's *On the Origin of Language*,[12] and even as late as his 1967 course on Heraclitus. Uexküll clearly struck a chord with Heidegger.

But even though Heidegger held Uexküll's thought to be of great importance to biology, it is more important for us to note where and how the two disagreed. The language that Heidegger employs to characterize Uexküll is unique in that it is always positive, but he is always at the same time offering a gentle reproach. In nearly every direct reference, Heidegger praises Uexküll as offering great insight into animal life, yet it is as though he is also disappointed with (and thus not too critical of) Uexküll. One slowly sees that Uexküll didn't go far enough, didn't think through his analyses enough, that he didn't, in other words, sufficiently radicalize his project, as seen here: "His investigations are very highly valued today, but they have not yet acquired the fundamental significance they could have if a more radical interpretation of the organism were developed on their basis" (GA29/30, 383/263). The ground is underfoot, but Uexküll remains steadfast rather than making the leap across the abyss that Heidegger finds between animals and humans. For this is ultimately the problem that Heidegger finds with Uexküll's thought. It is not that he doesn't offer a strong interpretation of animals, since he does so with his theory of the *Umwelt*. Nor is it the case that Uexküll succumbs to the criticisms brought against the others; he tends to avoid both mechanist and vitalist repercussions, and he doesn't characterize living things as entities present-at-hand. Within Heidegger's reading, Uexküll seems to have done nearly everything right, everything but make a proper transition from animal life to the existence of human Dasein:

> However, the whole approach does become philosophically problematic if we proceed to talk about the human world in the same manner. It is true that amongst the biologists Uexküll is the one who has repeatedly pointed out with the greatest emphasis that what the animal stands in relation to is given for it in a different way than it is for the human being. Yet this is precisely the place where the decisive problem lies concealed and demands to be exposed. (383/263–64)

The problem that Heidegger reveals is that although Uexküll offers insightful glimpses into the world of animals, and even though he points out a difference between animal and human worlds, he has not adequately described the essential manner of this relation underpinning the relation to world. Uexküll's vulnerability in Heidegger's reading is that he does not really offer an account of Dasein's relation to the world, one characterized by being able to "apprehend something *as* something, something *as* a being" (384/264). The absence of the as-structure or, more simply, the lack of any sustained discussion on the world of human beings, has led to Uexküll's inevitable fall.

No matter how great his descriptions of animal worlds might be, his failure to consider the world in terms of Dasein overshadows his entire project. Why? Because, as Heidegger will argue, in not conceptualizing human worldhood Uexküll has not really even begun to think through the concept of world itself:

> ... transcending any supposedly terminological issue, it becomes a fundamental question whether we should talk of a world [*einer Welt*] of the animal—of an environing world or even of an inner world [*Umwelt und gar Innenwelt*]—or whether we do not have to determine that which the animal stands in relation to in another way. Yet for a variety of reasons this can only be done if we take the concept of world as our guiding thread. (384/264)

This issue elaborates on the note that Heidegger makes in *Being and Time*, where he comments on the everyday parlance of 'having an environment' that he affiliates with the state of biology ever since Baer. The problem, as he sees it, lies in what is meant by 'having.' Unless the 'having' is clarified, the world or an environment is baseless. In short, we have the mark of a fundamental discrepancy in understanding the relation to the world, to the extent that Heidegger, despite his praise, in fact pulls out the ground from beneath Uexküll. As it turns out, Uexküll never really appreciated the concept of world insofar as he has not dealt with human existence. Heidegger is sure to clarify that it is neither a qualitative nor a quantitative difference, or even a terminological issue. Again, it comes down to the precise understanding of this relation between the organism and its 'world' or 'environment.' If the relation is not of a certain 'type' or, as we will shortly see, of a certain leeway in being open, then it may not even be appropriate to speak of 'world' in such circumstances. While this may at first glance seem to be exactly the kind of terminological issue that Heidegger says it is not—surely, one is tempted to say, we are just playing on the term "world"?—this first impression merely underscores how the decisive problem has been concealed. The ontological status of the animal's environment is at stake.

In this manner, we discover the guiding thread of Heidegger's engagement with contemporary biology. To enter this discussion, Heidegger argues that more than anything else it is the concept of world that provides the key. It may be the case that he grants himself a bit of leeway too, since he is not really concerned with animals per se, but rather with what they provide in a comparative look at the concept of world. It is the question of "what is world?" that he is pursuing in this lecture course and it just so happens that to get to this question he finds himself asking questions such as "what is the essence of life?," "what is the essence of animality?," and "what is the essence of the organism?" However, this ulterior motive by no means diminishes his fascinating and at times perplexing encounter with animal life, nor does his focus on world weaken his analyses of animals and organisms. By following Uexküll's lead, this engagment with animal being instead deepens his reading.

Even though Heidegger in the end finds Uexküll's biology fraught with difficulty, it is no less important. For in a way, this is the same diagnosis that Heidegger gives in his other readings of the philosophical tradition. Kant, for instance, "shrinks back" before the problem of being (GA2, 23/45), Schelling becomes "stranded" in the face of the *Abgrund* (GA42, 3), while Nietzsche is said to have "broken down" for the same reason (ibid.). All were entertaining the advent of something altogether new; all came up just a little short in their analyses. Though he does not belong to the same pantheon as these others, Uexküll falls short in his own way too. And yet, his look at the lives of organisms from the basis of the environment is the closest that Heidegger comes to finding a sympathetic view in the biological domain. He has his thread—the concept of world—and he finds its appearance in Uexküll's thought, more so than in any other biologist. The animal's relation to world awaits to be further radicalized.

THREE PATHS TO THE WORLD

It is fairly safe to say that insofar as Heidegger is concerned with the question of being, so too is he concerned with the understanding of world. The relation between human Dasein and world is an essential one, which Heidegger begins to capture in his hyphenated neologism "being-in-the-world" (GA2, 53/78). But what is the nature of this *unitary* phenomenon that is structurally a whole? Many questions surround the guiding thread of his analyses. What does it mean to be "in the world"? And what is this "as a whole" that Heidegger calls world? How does it reveal itself? How does this relation, furthermore, ground a priori the understanding of Dasein's being? And, just as important, how does this relation between being and world manifest itself in the case of other living things besides Dasein, such as animals, plants, and rocks? For this is precisely the problem that Heidegger found underlying Uexküll's treatment of the world or environment of animals: that it may be the case that the world does not pertain to all living beings equally, if it pertains at all.

Early in the second part of his lectures on *The Fundamental Concepts of Metaphysics*, when just commencing a more thorough questioning of "what is world?," Heidegger reminds his audience of three directions that could be pursued to reach a sufficient answer. One could ask after the term "world" itself and inquire into its etymology and the history of its use. Alternatively, one could ask about the world in its everyday sense, such as he does in his analytic of Dasein throughout *Being and Time*. A third path one might take, and introduced here in his writings for the first time, would be to follow a comparative path by asking about the world not just of humans, but of other organisms as well. Since it is the ontological nature of the relations between living beings and the world that we are inquiring after, it would be helpful to have a quick look at each of Heidegger's three paths toward understanding the concept of world, even if it is primarily the third comparative path that will prove most beneficial to our reading.

The History of the Concept of 'World'

Heidegger admits in his 1929–1930 lectures that he only offers a general glimpse of the historical stages of the concept of world. These stages can be found in various works, though perhaps most succinctly in his 1928 lecture course *The Metaphysical Foundations of Logic* and 1929 essay "On the Essence of Ground." In the latter essay, he likewise states that this type of characterization will undoubtedly leave "certain gaps" as he follows the term from Heraclitus' concept of 'χόσμος' through St. Paul and St. John, to the Latin '*mundus*' of St. Augustine and the Scholastics, and to the more modern conception of '*Welt*' offered by Kant (GA9, 38–3/111–21). In presenting the history of the concept 'world' in stages, Heidegger primarily emphasizes the "exterior" connotation rendered by the term: the world is taken to be something that lies 'out there' as the totality of beings in which we are a part. The idea of the world as something akin to a "blind mass of being," as Löwith explains (37), is contrary to Heidegger's conceptualization of the world as a way of being that opens onto the clearing of being itself.

Departing from the Greek concept of cosmos, Heidegger describes the world as a "mode of being" (GA9, 39/112). It is differentiated from the Greek *phusis*, or nature, in that the cosmos embraces all that *is*, particularly in a manner that emphasizes a prior totality and in that it develops around human Dasein.[13] But despite the importance of the Greek perspective, it is "incontestable" that the cosmos still refers to the totality of beings, thus signaling a certain deficiency. With the advance of Christian thought, one discovers a "new ontic understanding of existence that irrupted in Christianity" (GA9, 40/112). Chief among these perpetrators were St. Paul and St. John, each of whom contributed to an increasing focus on the human being within the world. The cosmos is no longer a 'cosmic' mode of being but an anthropological framing of the world wherein human Dasein is "removed from God" (40/112–13), alienated in his and her worldly condition. The world is now more of a human affair than the totality of things. Both of these characterizations are furthered by St. Augustine's use of *"mundus"* that reiterates our separation from God. But Augustine employs *mundus* in a dual sense: on the one hand, to denote "the whole of created beings" in a universal sense, and, on the other, to imply a carnal, earthy sense of dwelling in the world of flesh (40-41/113). The former use of *mundus* becomes adopted within Scholastic metaphysics in which the communion with God, as distinct from the world, actually becomes a means for knowing the world. "Here world is equated with the totality of what is present at hand, namely, in the sense of *ens creatum*. This entails, however, that our conception of the concept of world is dependent upon an understanding of the essence and possibility of proofs of God" (42/114). Within both the Christian and Greek traditions, the world transcends the concept of nature regardless of whether it refers to the totality of beings or the human condition. Löwith, in his reading of Heidegger, emphasizes that nature is subordinated to the world because it cannot address the ontological character of being that Heidegger is after. This account of the world underscores not only the ontological orientation of Heidegger's thought but also the implicit suppression of nature. Nature is just one kind of being among others, whereas the world looks to be the privileged source onto the opening of being.

The final stage that Heidegger recounts is the contribution made by Kant. After briefly addressing Kant's early 1770 Dissertation, in which Kant largely remains indebted to the metaphysical definition of world as the totality of beings, Heidegger notes that a particular problem is at work leading up to the *Critique of Pure Reason*. The issue comes down to an interpretation of finitude. Heidegger first summarizes Kant's thought in a series of three questions: "(1) *To what* does the totality represented under the title 'world' relate, and to what alone can it relate? (2) *What* is accordingly represented in the concept of world? (3) What *character* does this *representing* of such totality have; i.e., what is the conceptual structure of the *concept* of world as such?" (GA9, 44/116). These questions are especially noteworthy insofar as they exemplify the direction and focus of Heidegger's reading. Kant offers a novel approach in that the concept of world undergoes a significant change and takes on new meaning. What emerges is that the world continues to be related to "finite" things—as it had been throughout the history presented here—but the relation to finitude is understood differently. "The finitude of things present at hand," Heidegger writes, "is not determined by way of an ontic demonstration of their having been created by God, but is interpreted with regard to the fact that these things exist for a finite knowing, and with regard to the extent to which they are possible objects for such knowing, i.e., for a knowing that must first of all let them be *given* to it as things that are already present at hand" (44/116). By reframing this relation to finite things, Kant demonstrates how things are given to knowing in a particular fashion:

beings that are knowable are knowable only as "appearances," as opposed to an absolute understanding of their being as "things in themselves."

Kant's theory therefore underscores the fragility of an absolute epistemological relation to the world. But not only is the world questionable as something knowable in itself, it may also be knowable in different ways, as Uexküll demonstrates well. If the world is only capable of becoming a totality or unity in its appearance *for* us, then this "*unity of appearances . . .* is at all times *conditioned* and in principle fundamentally incomplete" (45/117). At best, then, we can have an idea of the totality of beings, but such an idea only carries representative value. And yet, Heidegger remarks on "the more originary ontological interpretation of the concept of world" that emerges out of Kant's transcendental ideal. Even though the world may only be accessible via its representation as appearance, this concept of world adheres to, but also transforms, the earlier notion of totality. The world, as totality, is now defined by Kant as "the sum-total of all appearances" (46/118). Captured in this turn of phrase is Kant's appeal to our human experience of this totality, a totality that may be conditional and incomplete, yet is nevertheless 'complete' in experiential life. Heidegger writes: "World as an idea is indeed transcendent, it *surpasses* appearances, and in such a way that as *their* totality it precisely *relates back* to them" (48/119). The totality of world is achieved through this reflexivity or, otherwise said, in the relation between "the possibility of experience" and "the transcendent ideal." This offers an explanation for why there are no real gaps in the world of experience, sections where the world leaks out of the enclosed totality. This totality, Heidegger restates, is found in the finitude of being human.

After reviewing these stages, however, the concept of world is still deemed unsatisfactory. Neither this interpretation of world, nor Kant's version in the *Critique of Pure Reason*, nor any of the other stages that Heidegger recounts in his historical overview. In addition to the aforementioned stages, we could also include Descartes' world as *res extensa* found in *Being and Time*, world in relation to Leibniz's monads in *The Metaphysical Foundations of Logic*, as well as many other accounts of world that Heidegger treats, right through to his critique of the *Weltanschauung* of his day. They all helpfully appeal to a relational structure, but none of them do so in an adequate manner. To quote at length, Heidegger concludes:

> what is metaphysically essential in the more or less clearly highlighted meaning of χόσμος, *mundus*, world, lies in the fact that it is directed toward an interpretation of human existence [*Dasein*] *in its relation to beings as a whole*. Yet for reasons that we cannot discuss here, the development of the concept of world first encounters *that* meaning according to which it characterizes

the "how" of beings as a whole, and in such a way that their *relation to* Dasein is at first understood only in an indeterminate manner. World belongs to a *relational* structure distinctive of Dasein as such, a structure that we called being-in-the-world. (51–52/121)

The structure of being-in-the-world that offers us an ontological grasp of the world's unity is found in Being and Time. As we have begun to see in Heidegger's reading of Uexküll, the concept of world is distinctive to Dasein's relational structure. This claim is reiterated in the passage just cited; it only remains to be illustrated. This everyday understanding of the world, one that is so familiar to everyone of us that it has for this very reason eluded our attention, is the second path in Heidegger's account of the world.

The Everyday Way of Being-in-the-World

Heidegger's central claim in the opening chapters of Being and Time is that all claims of knowing the world is in fact founded on a more prior understanding of being-in-the-world. In order to have an epistemology or metaphysics of things, the world itself must first be interpreted as the horizon in which everyday life plays out. Unless this happens, all other claims are ontologically groundless, as we have seen in the preceding characterizations of world throughout the history of philosophy. All lay claim to some relational structure of knowing the world, but do so in such a way that the world of our everyday activities remains obscured. Only within our ordinary daily dealings—from cooking a meal to talking with a friend—may the world be revealed in its foundational character.

This is the background to his statement in the 1929–1930 course that "I took my departure from what lies to hand in the everyday realm, from those things that we use and pursue, indeed in such a way that we do not really know of the peculiar character proper to such activity at all" (262/177). However, this does not mean, as he sarcastically remarks, that the world is revealed because "[man] knows how to handle knives and forks or use the tram." As we have already noted, Heidegger has no use for anthropological statements, such as the observation that humans are tool-users and the harbingers of culture. The manipulation of tools is not sufficient to light upon the essence of either Dasein or the world. And yet, tool use is nevertheless an access point for Heidegger's discussion of being-in-the-world. This explains the emphatic tone of his clarification of "equipment" (*Zeug* is also translated as "tool" or "useful things"). The everyday character of the world that is appealed to is one that aims toward interpreting *how* we encounter things in daily dealings. So this does mean a discussion of knives, hammers, and trams. But contrary to the tradition that Heidegger critiques, one cannot begin by making epistemological claims about the knife or fork and how such things are a part of the human world. For example, we learn little by noting that a utensil is made out of steel and wood and devised for stabbing or cutting food. Though these claims may be true about a knife, such a definition rests upon the prior realization that one encounters a utensil in this way only because it first appears within a world already at hand. It is, in other words, not the things as objective entities that interests Heidegger, but "rather a determination of the structure of the Being which entities possess" (GA2, 67/96).

The distinction that Heidegger makes in the understanding of things is his famous and well-traversed concepts of "readiness-to-hand" (*Zuhandenheit*) and "presence-at-hand" (*Vorhandenheit*). In order to broach being-in-the-world, Heidegger begins with commonplace things (e.g., knives, door handles, hammers) and how they appear in our midst. The distinctive feature of these things is that they do not originally appear as objects known within the theoretical purview of one's everyday existence. It is true that one might and can grasp a door handle and think 'This is a handle that I grasp and which allows this door to open such that I might enter the other room.' But we don't do this; we simply can't live this way. Rather one simply grasps the handle and opens the door without thinking twice about it. This, incidentally, is what is wrong with many cognitive models in psychology that endlessly break down human behavior into functional bits (e.g., see handle, extend arm to handle, grasp handle, turn handle, pull, etc.). Merleau-Ponty in particular will critique a similar form of psychological reductionism by noting that human behavior is constituted by the whole body in the fluidity of its movements. If I want to kick a soccer ball, for example, I don't think of all the steps required to kick the ball. Should I do so, my opponent would surely have already taken the ball away before I had a chance to kick it for myself! The body acts as a whole in virtue of always already being in a world. Everyday examples such as this one multiply innumerably over the course of everyday affairs. And this is precisely Heidegger's point: the world is not there before one's theoretical and objective glance, but is always already fundamentally there in how one exists. To conceptualize this original encounter with things Heidegger coins the term "ready-to-hand." Things are ready-to-hand insofar as they are worked, used, manipulated, and handled, without necessarily giving them any thought to their specific identity: "The peculiarity of what is proximally ready-to-hand is that, in its readiness-to-hand, it must, as it were, withdraw in order to be ready-to-hand quite authentically" (GA2, 69/99). This withdrawal is what lends things their stature as pretheoretical, or, as Merleau-Ponty will say, prereflective.

In contrast to encountering things as ready-to-hand, we also come across things that simply malfunction. The door handle is jarred and doesn't turn, your pen runs out of ink, or the subway is delayed on route to work. When an entity is torn away from the "totality of equipment," the thing suddenly becomes present-at-hand. The door handle is suddenly there as a door handle, the pen is suddenly a pen to thought as opposed to an implement that was being used in writing notes. The present-at-hand refers to one and the same entity as the ready-to-hand; only what is decisive in each case is that the present-at-hand refers to a thing once it has been ripped out of the fabric of everyday life, out of the otherwise "concernful absorption [besorgenden Aufgehens]" (71/101) that characterizes Dasein's being-in-the-world.[14] But once the theoretical guise is adopted, once the readiness-to-hand of the pen becomes present-at-hand, Dasein is no longer simply in-the-world of the ready-to-hand, since a reflective stance has been assumed and the pen (and thus the world) becomes an object for Dasein. In this case, the world is reduced to the ontic accumulation of entities present-at-hand.

In contrast, Heidegger's sense of the world originally arises out of the everyday mode of being absorbed in the familiar domain of one's daily dealings. I will have to return to the concept of Dasein's absorption later in the next chapter when it will be seen in comparison to the animal's relation toward the world. Within the parameters of this second path, however, it is the totality of the everyday world that concerns us: "The context of equipment is lit up, not as something never seen before, but as a totality sighted beforehand in circumspection. With this totality, however, the world announces itself" (75/105). The world, in other words, is not something present-at-hand like an object sitting before an observer waiting to be discovered and known. The world, rather, is the hidden context or horizon for everything we do: "In this totality of involvements which has been discovered beforehand, there lurks [birgt] an ontological relationship to world" (85/118). This sense of totality will become synonymous with the expression of the world "as a whole [im Ganzen]" (GA29/30, 251/169), a formulation that Heidegger is still clarifying in his 1929–1930 lectures.

An analysis of all the characteristics that Heidegger employs to give an account of the world would be too exhaustive here. The trick will be to later show how things reveal themselves, break open in their being, without however becoming present-at-hand. That is, how—and, in the case of animals, whether—they show themselves in their being without becoming torn out of the totality in this process. Things do not simply show themselves on their own, but do so in particular relation to human Dasein. Is it the case that the world opens up to animals in the same way? It is already suggested that this is not the case. Among many other features, involvement with the world relies on letting things be and having the presence of self in this relation, leaving things untarnished in the state of the ready-to-hand. Things are discovered out of this totality. Or indeed it is the totality itself of which we speak when Heidegger says the worldhood of world is "the being of that ontical condition which makes it possible for entities within-the-world to be discovered at all" (GA2, 88/121). But do not animals also have this capacity to exist in relation to the world?

The world, suffice it to say, is not simply an empty horizon awaiting things, people, atoms, and so on, to populate it. Further, just as the world is not a metaphorical container filled with 'stuff,' nor are things and entities mere stuff that fills the world. At issue then is the relational structure between Dasein and the world such that the world becomes disclosed in its worldhood. Heidegger writes: "World belongs to a *relational* structure distinctive of Dasein as such, a structure that we called being-in-the-world" (GA9, 52/121). We have already seen in his reading of Uexküll that Heidegger accords the world a specific relation with Dasein, a relation that he is not yet willing to grant to other animals, at least not without further thought. Underlying this claim is the belief that there must be something special to Dasein that allows the world to be and disclose itself, which may or may not be applicable to animals as well.

The phenomenon of world in the everyday dealings of human Dasein has at least been opened, even if it has not been comprehensively reviewed. Sometime during the years of the writing of *Being and Time* and the 1929–1930 lectures, Heidegger clearly faced a problem. In order to authoritatively state that Dasein is distinctive in its relation to world—that Dasein even 'has' a world, while other living and nonliving things may not—would in the end require an ontological analysis of other beings. For even the most exhaustive analysis of Dasein's relation to world would not be able to evade the bias attributed to Dasein. How do we know, one might ask, that Dasein, and Dasein alone, has the capacity for disclosing the worldhood of world? Animals live in the world and make use of things too, don't they? Do they not also have a relation to world? And what about plants? And rocks? Furthermore, how is this relation toward the world *the* indication that defines a living thing as an organism or as an existing being? To better clarify the concept of world, Heidegger realizes that it would be advantageous to offer a comparative examination, the third path to the world, which we discover in the 1929–1930 course. He had already given a fair bit of treatment to Dasein's being-in-the-world throughout the 1920s, but by the end of the decade he clearly realizes that an approach comparing different living and nonliving things might better explain the world and what is so distinctive about Dasein. The time for animals had arrived.

A Comparative Examination of Worlds

Heidegger admits that there are other paths that one could possibly follow toward a clarification of world. But he neither lists what these paths might be, nor does he go into any further detail on this possibility. Instead his attention turns to the more immediate concern of a comparative examination to help answer his guiding question. Now why, one might wonder, does he concentrate his focus on this third path, a path that will take him into the difficult terrain of animal life and material things? This direction is all the more bewildering given his propensity for specifically *not* dealing with seemingly anthropological and biological themes. We must recall that his primary focus is on the concept of the world, especially in its relation to finitude and temporality. His focus is not on animal life specifically; it is of interest only insofar as it fits into the bigger picture. Here is how Heidegger broaches the topic:

> Man has world. But then what about the other beings which, like man, are also part of the world: the animals and plants, the material things like the stone, for example? Are they merely parts of the world, as distinct from man who in addition *has* world? Or does the animal too have world, and if so, in what way? In the same way as man, or in some other way? And how would we grasp this otherness? And what about the stone? (GA29/30, 263/177)

Indeed, what about these other beings? All are good questions. And yet, as quickly as they are posed, they are just as quickly shuffled into admittedly "crude" theses. Before humans, animals, and stones have a chance to settle to into a confusion of blurry boundaries, they are divided into firm distinctions: the stone (and all other material objects) is *worldless* (*weltlos*), the animal is *poor in world* (*weltarm*), and humans are *world-forming* (*weltbildend*). As for the neglected realm of plants, which Heidegger largely forgoes, one can find an instance of his reading in the 1939 essay "On the Essence and Concept of Φύσις in Aristotle's *Physics* B, 1" (GA9, 324–25/195). Another characterization of plants is issued in the 1946 essay "The Origin of the Work of Art," where plants are more or less equivalent with animals, at least as concerns world: "A stone is worldless. Plant and animal likewise have no world; but they belong to the covert throng of a surrounding (*Umgebung*) into which they are linked. The peasant woman, on the other hand, has a world because she dwells in the openness of beings" (BW, 170). Insofar as plants are organisms too—Heidegger does refer to uni- and multicellular life, such as algae—there is no reason to think that Heidegger would characterize plants any differently.[15]

The significant picture of each characterization is that human beings are offset from the rest of nature. Nowhere is this stated more clearly than in Heidegger's "Introduction to 'What Is Metaphysics?'" which he wrote in 1949 for the fifth edition of his 1929 lecture. Here in this introduction he leaves no doubt as to where humans stand in relation to everything else.

> The being that exists is the human being. The human being alone exists. Rocks are, but they do not exist. Trees are, but they do not exist. Horses are, but they do not exist. Angels are, but they do not exist. God is, but he does not exist. The proposition "the human being alone exists" does not at all mean that the human being alone is a real being while all other beings are unreal and mere appearances or human representations. The proposition "the human being exists" means: the human being is that being whose being is distinguished by an open standing that stands in the unconcealedness of being, proceeding from being, in being. (GA9, 204/284)

The great chain of being, so decisive in previous ontologies, has been torn open, and it is not God that stands outside in the open but human beings. Humans alone *exist*. Inasmuch as human existence is framed first through its transcendent condition as being-in-the-world, it is more necessary than ever to consider the three theses that got the discussion rolling. And while human existence is not our immediate interest, it is through the 1929–1930 comparative analysis that Heidegger can eventually state so matter-of-factly, as he does in the previous quote, that humans are on one side of a line, while everything else is on the other. But in order to arrive at this situation, he needs to initially settle the muddle of distinctions that one observes in the late 1920s. Therefore, with a set of theses—mere hypotheses really—we are prepared to assay Heidegger's comparative analysis of world. What was once passed off as irrelevant has now reappeared. Even if animal life is still only of secondary importance to the key role of Dasein's world, the animals threaten to steal center stage. They are right in the middle of Heidegger's encounter with the world of biology.

CHAPTER 3

Disruptive Behavior

Heidegger and the Captivated Animal

In a recent introduction to the philosophy of cognitive science, Andy Clark begins by asking if there are any strong distinctions among the elements of nature. "What," he asks, "distinguishes cat from rock, and (perhaps) person from cat?"[1] The examples—a rock, a cat, and himself, a human being—are interesting insofar as they parallel Heidegger's own theses regarding the natural world, but the differences between them are even more telling. After briefly recounting a day in the life of each, Clark arrives at three types of phenomena that may distinguish one type of being from the other: (1) experiential "feelings" such as hunger or desire; (2) thoughts and reasons; and (3) the "meta-flow" of thoughts about thoughts. Since Clark is entertaining theories of cognition, these enumerations may not be surprising, but in spite of this they do illustrate a particular angle in the categorization of nature. Just like in Heidegger, the natural order is divided between rock, animal, and human, but here the attention is placed on "mentalistic discourse," on the ability of beliefs, thoughts, and desires to explain a living being's actions. The mental world, however, is no longer the incorporeal and spiritual life of the mind; here mindfulness and mental life is no more than the orchestration of a material brain, the corporeal body, and its proximity to a physical world.

Clark's materialist monism helpfully positions our own look at Heidegger's unique approach. Even though Clark frames his account with respect to the concept of world—and he does so with awareness of Heidegger, Merleau-Ponty, and in his book *Being There*, with reference to Uexküll as well—one still gets the sense that the emphasis on mental life misses Heidegger's more rudimentary ontological priority. Of course, Clark and Heidegger are working on different projects, but this is what makes their comparisons between rock, animal, and human all the more engaging. With Clark, all elements of nature are part of "the whirr and buzz of well-orchestrated matter" (5), including especially mental life. Thus, the ordeal is to decipher the emergence of mental life out of the otherwise brute physical matter. Heidegger, on the other hand, is not concerned with the thoughts, feelings, beliefs, or desires of rocks, animals, or humans, but with how they relate to their environments through their behavioral being. Conscious life is clearly not the issue here. Prior to any such effort, the ontological status of living beings must first be settled. Rather than asking questions about mental life, Heidegger wonders if all beings relate to the world in the same way. Is the being of the animal or rock similar to being human? In a way, Clark already assumes the distinction between rock, animal, and human before beginning his inquiry. Though we could say the same of Heidegger, his ontological emphasis on behavior sets the foundation one step further: not a distinction between beings, but between ways of being.

On a historical note, it is particularly striking that Heidegger formulates his three hypotheses concerning rock, animal, and human just a few years after the appearance of some notable publications in philosophical anthropology that implicitly critique Heidegger's account of human Dasein. Still early in his 1929–1930 lectures, Heidegger acknowledges some recent publications on the contemporary character of the modern European (104–107/69–71). Not included, however, are Max Scheler's *Man's Place in Nature* and Helmuth Plessner's *Die Stufen des Organischen und der Mensch* that, though each inspired by Heidegger, saw his account of human Dasein in *Being and Time* as too devoid of life. It is as if Dasein were a being seemingly set apart from the rest of nature, a critique that we have also observed in Karl Löwith and Hans Jonas. In their respective writings, both Scheler and Plessner sought to 'return' humans back into the context of nature without necessarily naturalizing them, and did so in keeping with more traditional philosophical anthropology.[2] Humans, even if distinct, were placed back squarely in the midst of nature.

Despite the lack of acknowledgment of these works, Heidegger was certainly familiar with them. A reply was in order, something more needed to be said, and what we discover in the 1929–1930 lectures is a more concerted and explicit means to address not only human Dasein's relation to the world—which is not the same as a 'return' to nature—but also that of other living beings. Though it is clear that Heidegger held Scheler in particular in high esteem, as seen in many instances, such as his memoriam to Scheler's sudden death in the midst of his 1928 course (GA26, 62–64/50–52), he was also critical of his anthropological approach. As a result, the 1929–1930 lectures could be read, as David Krell has expressed, as an effort to both acknowledge and overcome his "fraternal rival" in a final and decisive way

(82). The trio of theses thus represents, on the one hand, a means of settling, for the "first time," a fundamental problem of metaphysics, namely, the world (GA29/30, 264/178), and, on the other, it responds to the current talk of the day that regards human existence as just another part of nature, not to mention the suggestion that animals 'have' environments too.

To pursue these distinctions, it is not enough for Heidegger to remain content with the traditionally drawn lines that demarcate humans as "rational living beings" compared with the "nonrational" nature of all other nonhuman animals. Nor is it suitable to observe that humans have descended from the ape, a distinction that draws only a fuzzy evolutionary line and speaks little to the ontological clarification of these two roughly delimited groups. Rather, Heidegger notes, "it means finding out what constitutes the *essence of the animality* of the animal and the *essence of the humanity* of man," which, more important, necessitates the even more general question of "what constitutes the *living character of a living being*" (265/179). An initial observation is that life can be explained by means of human Dasein—namely, as a certain lack of what Dasein fundamentally is—but Heidegger is clear that this relation is not reciprocal: Dasein cannot be explained by means of life because this would amount to suggesting that Dasein is merely life plus an additional something else. A similar dynamic will also ultimately hold of the animal-human relation: the animal can be explained as poor-in-world in comparison to humans, who are world-forming in their capacity to *have* world, but the comparison cannot be reversed. Humans cannot be referred back to animals because of the "abyss" that separates the two, in that this distinction is for Heidegger "a statement of essence." The claim that life only pertains to animals and plants—Heidegger wishes "to restore autonomy to 'life,' as the *specific manner of being pertaining to animal and plant*" (277/188)—rests on the assertion that the matter is primarily one of essence. To speak of animals requires that one say something about their essence. To carry this thought further, Heidegger makes the more controversial statement that the essence of animals "holds true for all animals because it is a statement of essence." Essence has a "universal validity," and thus what is said of the lizard, bee, lion, or eagle, also holds true of all reptiles, insects, mammals, or birds, as well as "non-articulated creatures, unicellular animals like amoebae, infusoria, sea urchins and the like—*all* animals, *every* animal" (274–75/186).

A question that we will have to keep in mind while looking at Heidegger's theses is whether or not his distinction between animals and humans is obscured by his lumping all animals together in one category. It is toward such a statement that Jacques Derrida directs one of his criticisms in *Of Spirit*, and more recently in his essay "The Animal that Therefore I Am (More to Follow)." In *Of Spirit*, Derrida concludes his brief reading of the 1929–1930 course with a few open-ended remarks, including the question of whether Heidegger's second thesis on animality might not be compromised due to its composition: "Compromised, rather, by a *thesis* on animality which presupposes—this is the irreducible and I believe dogmatic hypothesis of the thesis—that there is one thing, one domain, one homogeneous type of entity, which is called animality *in general*, for which any example would do the job" (57).[3] There is no obscurity in Heidegger's use of the catch-all terms "animal" and "animality" to represent all animals, aside from humans of course. Indeed, in Heidegger's use of "animal," it is not only doubtful but clearly denied that humans are animals themselves. Conceptually, they are worlds apart. This might at first seem peculiar, if not outrightly false. Derrida, for one, remains unconvinced that all animals deserve to be categorized under the one concept of "animal." From a biological standpoint, there can really be no question that human beings belong to the animal kingdom. But here we are slipping dangerously between the border of biology and ontology. Appealing to biology may also have somewhat of a paradoxical effect in that it can reinforce Heidegger's usage of "animal" in the plural. For example, it is common to find in genetic research the use of something like a "model organism" that can represent many if not all other animals, including humans. In previous years, the standard model organism had been the fruit fly (*Drosophila melanogaster*), and more recently the roundworm (*c. elegans*) has stood in for the archetypal animal due to its sequenced DNA and defined developmental pattern. In psychology, too, we can look to the infamous and, I would add, highly problematic use of rats and chimpanzees to map and understand the behavioral and neurological functioning of humans. The use of such model organisms is a separate issue entirely from what Heidegger means by the essence of animality, but they do provide something of a glimpse at why it might be problematic to lump all animals together.

If we hesitantly condone for the moment that of all biological beings, humans stand apart, then what is the essence of animality? Again, this depends on the conceptual issue of the environment and world. The kind and manner of access that an animal has to something like a world makes all the difference, and even forms a provisional definition of world: "Let us provisionally define world as those beings which are in each case accessible [*zugänglich*] and may be dealt [*Umgang*] with, accessible in such a way that dealing with such beings is possible or necessary for the kind of being pertaining to a particular being" (GA29/30. 290/196). This is a very rough formulation since Heidegger will have occasion to question and refine this understanding of world over the course of his ensuing lectures. But what this statement does do, for the moment at any rate, is rule out the possibility for material entities to have something like a world. Before reaching the being of the animal, we can first cast aside the flat, one-dimensional, and worldless rock.

THE WORLDLESS STONE

Heidegger embraces the stone as representative of the worldlessness of material things. "The stone," Heidegger preemptively writes, "is worldless, it is without world, it has no world" (GA29/30, 289/196). Pretty definitive, but is this the end of the story? Almost. The stone can be said to have no world because in the manner of its being a stone, it is accorded the impossibility of ever accessing or relating to the things around it. Heidegger explores different possibilities by which a stone might be said to relate to other things: the stone lies on the earth, it has a lizard lying on it, it is thrown into a ditch, it sinks in water, it has pressure exerted on it by the force of gravity. In each case, the earth, the lizard, the water, force, and so on, is not given to the stone in any mode of accessibility. But the stone, one might object, surely has contact with each of these other entities. Doesn't the stone 'touch' the earth? When the sculptor takes her chisel to a piece of marble, isn't there a relation in the chisel striking and chipping away the stone? Or isn't there a visible relation between stone and water, as seen in the case of pebbles on a beach or in the hoodoo rock formations of southern Alberta? More than others, Graham Harman has questioned Heidegger on his ability to think about objects and their relations to one another. His "guerilla metaphysics" seeks to enliven the metaphysics of the event arising out of the contact between things. Harman, I think, rightfully pushes Heidegger's thinking of ontic entities, but Heidegger, meanwhile, won't have any of it. It is undeniable that stones come into contact with other entities (Heidegger never denies this), but contact itself is not enough to constitute a relation, nor is it therefore enough to have world. At stake in the "relation" is something more than simple contact between two different material entities (e.g., a stone hitting earth, a ladder leaning against a brick house, a book resting on a table). It is not merely because such contact is incidental (whether such contact is incidental is not of much concern), but because a relation implies a more significant access that opens the possibility of a world *for* the material thing. Even as late as 1967, in his course on *Heraclitus*, Heidegger still expresses that "the concept of ontic proximity is difficult. There is also an ontic proximity between the glass and the book here on the table" (GA15, 232/144). But this is not necessarily an ontological proximity.

Instead, Heidegger explains that what is required to have world is the enigmatic "as" that opens, manifests, and otherwise makes accessible the essence of a being *as* the being it is. The earth, the lizard, and the water remain inaccessible to the stone, in their being, as the beings that they are. So for this reason alone, Heidegger quickly concludes that the stone is worldless, it has no world, precisely because the stone—and any other material thing—is incapable of having any access to something else: "it has no access to beings (*as* beings) amongst which this particular being with this specific manner of being is" (290/197). But this still doesn't capture the full extent of the stone's worldlessness. In not having world, Heidegger writes, "the stone *cannot even be deprived* of something like world" (289/196) because the world is not accessible in the stone's specific kind of being. Describing the stone as neither having nor not-having world is informative because the world is not even there as a possibility. It is not a case of the absence or a lack of world, since such a conception implies the possibility of having the presence of a world. The stone is fundamentally denied any possibility of relation, and thus so too of world.

This reading finds some support from a somewhat different angle proposed by Jean-Paul Sartre in his account of the being-in-itself of material things. Without involving ourselves too much, Sartre offers an interesting example in *Being and Nothingness* on how material things 'relate' with one another in the absence of any conscious perception. His example comes early in the text as he describes the destruction of a building after a natural disaster.

> In a sense, certainly, man is the only being by whom a destruction can be accomplished. A geological plication or a storm does not destroy—or at least they do not destroy *directly*; they merely modify the distribution of masses of beings. There is no *less* after the storm than before. There is *something else*. Even this expression is improper, for to posit otherness there must be a witness.... In the absence of this witness, there is being before as after the storm—that is all. (8)

Among other things, this example hits on the notion that there is neither more nor less of the being of 'rock' before or after the destruction. Any such determination requires a "witness" to perceive difference; a difference in relation is not perceived by the rocks themselves. Without the being-for-itself of human existence, Sartre will claim, all of being flows into an indecipherable being-in-itself. The rocks have no relation to their surrounding, since no difference exists in itself between the rocks as a building or as rubble. To note this difference, to witness a relation, a witness is needed (for our purposes, it may be interesting to note that the witness is assumed to be a human being, though we can imagine that an animal would also 'witness' a difference in material surroundings). Even though Sartre's focus emphasizes a bit too much the conscious perception of difference, his example is nevertheless a useful one. For Sartre, as it is for Heidegger, material things have no relation to other things. They may be acted on, but have no access to others. They merely are.

And yet, one would have to at least wonder about the nature of this relation. While Heidegger is quite clear that a material thing, in and of itself, does not even have the possibility of relation, is it not the case that a lizard relates to the rock beneath its body, or that a human relates to the rock that he skips across a lake? In either of these examples, it is still assumed that the stone does not have world because it does not reciprocate any relation to the lizard or human. But insofar as a relation is established between the lizard and rock or the human and rock, doesn't any such relation imply some form of reciprocity between the two (or three, or four . . .) things involved? And if this is the case, wouldn't such a relation necessarily involve all parties, including a relation on the part of the material thing(s)? Could we say that the rock is transformed in some still uncertain way through each relation, even if the possibility of a relation is still in question? Or is it simply the case that a relation can be only one-way? If this is the case, what kind of relation is this? Again, the issue comes back to the "having" of world, and the rock still does not have world within the parameters of Heidegger's analysis. Let us leave these questions aside for the moment, since Heidegger is quite clear on his position: there is no possibility for relation and thus no possibility of access to something like world. But while we pass them by for the time being, we will encounter different responses, and perhaps answers slightly more sympathetic toward material things, from both Merleau-Ponty and Deleuze. It is clear, however, that there is a difference between acknowledging that rocks may be present within the worlds of humans or animals and saying that rocks have a world for themselves.[4] The two claims are worlds apart. This description of the stone will prove to be a helpful juxtaposition for turning to the second thesis concerning the animal's connection to a world.

THE POOR ANIMAL

It is true that from the outset Heidegger must withhold world from the animal if he is to follow the same maxim that he has already established with respect to the stone. The animal, like the stone, does not have access to being as such. However, unlike the stone, animals do have some kind of relationship with things despite their inability to grasp things *as such*. A simple glance at any living being will demonstrate this. A dog eagerly chases after a squirrel, a moth is drawn toward a source of light, a bird feels the branch beneath its feet, an amoeba recoils from its encounter with another. Each demonstrates a relation, each some sort of access. But there is a disclaimer on the scope of these relations, for while Heidegger admits that animals have a "specific set of relationships," the mode of these relations is not without qualification.

At stake is the same issue as was the case with material things: whether animals have access to other beings *as* the beings they are, whether they have access to the question of the meaning of being as such. This "as-structure" is proving to be the make-or-break phenomenon that decides whether a given thing is capable of having a relationship with an other, and so too then of having world. In the case of animals, the nature of this relationality is much more difficult than the suspiciously easy case of material things. In what may turn out to be a case of foreshadowing, Heidegger characterizes this task of understanding animal relations as "infinitely difficult for us to grasp," for reasons soon to be seen. Part of the problem is that animals, unlike stones, have a certain access to things. The lizard, for example, searches out for the right stone to lie on, angling itself in a specific manner to the sun. The rock, in comparison, has no such relation to the lizard. But it is not just that the lizard has a relation to this rock (as opposed to another), and to the sun (as opposed to the shade, or a tree, or any other thing). Even though it is given that the possibility of a relation is present in the case of animals, it remains to be questioned what kind of relation this might be. What is the manner or 'how' of this relation? As seen in the example of the lizard, Heidegger explains that the lizard does not have access to the rock *as* a rock, despite the appearance of some relation: "When we say that the lizard is lying on the rock, we ought to cross out the word 'rock' in order to indicate that whatever the lizard is lying on is certainly given *in some way* for the lizard, and yet is not known to the lizard *as* a rock. If we cross out the word [rock] . . . we imply that whatever it is is not accessible to [the lizard] *as a being*" (GA29/30, 291–92/198). Derrida draws a compelling connection between this near erasure of the word 'rock' and Heidegger's later crossing through of being in his 1955 essay "On the Question of Being" (GA9, 238–39/310). The focus of Derrida's reading highlights how Heidegger mentions that we "ought" to cross through rock (~~rock~~), but that this erasure remains suspended ("ought," "if"), and therefore never fulfilled (*Of Spirit*, 52–54). The rock is both there and not there for the animal. It is present, but it is also cloaked in a near absence. Derrida's emphasis captures well the homogeneity between Heidegger's linguistic presentation of this idea (the almost crossing through of the rock) and his more pronounced argument concerning animals (the almost crossing through of a relation, of world). This linguistic play points toward a more important discovery, however: the rock is present to the lizard, but not *as* a rock. The rock is given "*in some way*," but not given as a rock as such.

This flexibility gives rise to the cautious description of the animal and world. If, as Heidegger notes, "we understand *world* as the *accessibility of beings*" (GA29/30, 292/198), then it seems that all animals, like the lizard sitting on the rock, are sitting on the figurative fence. They are caught in the middle, both as a natural entity between material things and human beings, and with respect to having and not-having a world. Subsequently, Heidegger claims that the "animal thus reveals itself as a being which *both has and does not have world*" (293/199), and that it is this manner of being that best typifies what he understands by the word "life." As a conditional summary still in need of clarification, we approach the being of the animal:

> The animal's *way of being* which we call *"life,"* is *not without access* to what is around it and about it, to that amongst which it appears as a living being. It is because of this that the claim arises that the animal has an environmental world [Umwelt] of its own within which it moves. Throughout the course of its life the animal is confined [*ist . . . eingesperrt*] to its environmental world, immured as it were within a fixed sphere that is incapable of further expansion or contraction. (292/198)

The direction of Heidegger's thought concerning the animal world is already evinced within this informative paragraph. The animal, Heidegger allows, has an environment; or, at the very least, he acknowledges that such a claim has arisen and that he is not presently objecting to it (not here at any rate). On this point Heidegger is clearly following the observations made by Uexküll, to whom we are indebted because "we have all become accustomed to talking about the *environmental world of the animal*" (284/192). And yet Heidegger is not particularly recognized for going along with the "accustomed" way of talking, so it should be with a bit of skepticism that we follow his adherence to the view that animals have an environment. His language already betrays his position; the conditions of this environmental world are limited because the animal is "confined" to its "fixed sphere." He is almost suggesting that animals do not have an environment at all, a remark that he will make explicit in his 1935 course, *Introduction to Metaphysics*. By then, "the animal has no world [Welt], nor any environment [Umwelt]" (GA40, 34/47). But we are not yet at this point. The animal is not without access either, so the possibility of world is still ambiguous.

The concept of having and not having a world clearly implies a certain relation between the animal and world, and to get to the bottom of the relation we need to go further into the essence of the animal. To state it upfront, Heidegger repeats consistently throughout the 1929–1930 course that the essence of the animal is captivation (*Benommenheit*) (348/239; 361/248;

409/282). Had this statement been made by a biologist, "captivation" would surely have left more than the animals dumbstruck. *Captivation is the essence of animals?!* Captivation? As it is, Heidegger's essential definition of animality seems strange enough. But this is only if we continue to think of the animal as an ontic being. The trick is in not getting caught thinking about the animal as a biological entity. Just as the essence of being human lies in existing in the world, the essence of animality lies in proximity to the thesis "poor in world." Heidegger is always aware that we might be susceptible to thinking of the animal in the wrong way, so he makes sure we are always on the right path:

> We might be tempted to fall back on the notion that φύσει-determined beings could be a kind that *make themselves*. So easily and spontaneously does this idea suggest itself that it has become normative for the interpretation of living nature in particular, as is shown by the fact that ever since modern thinking became dominant, a living being has been understood as an "organism." No doubt a good deal of time has yet to pass before we learn to see that the idea of "organism" and of the "organic" is a purely modern, mechanistic-technological concept, according to which "growing things" are interpreted as artifacts that make themselves. (GA9, 325/195)

This claim arises in his later 1939 essay "On the Essence and Concept of Φύσις in Aristotle's *Physics* B, 1," but it holds just as easily to his descriptions of the organism ten years earlier. In this same essay, Heidegger relates that this notion of the organism as a growing thing capable of making itself is what might lead to the utterly devastating technical feat of humanity *"producing itself technologically."* In our current age, where cloning is now common in so-called higher animals such as pigs and horses (and humans are not too distant), one need not wonder too much at why this definition of the organism may be problematic. Needless to say, interpreting the organism as simply a self-generative being does not delve deep enough to speak to what is fundamental about the being of living things. On this point, Heidegger finds the concept of "organism," at least in its modern usage, highly problematic, which in part explains his own vacillation in making use of the concept. "Organism" will tend to refer to the biological entity, and therefore hold for all living beings, while "animal" refers to a *"more originary structure"* (GA29/30, 341/234) characteristic of organic life. It just so happens, rather uncoincidentally, that both animal and organism share the essential mode of being captivated (376/258). As we shall see, Deleuze too is highly concerned with the concept "organism," though for slightly different reasons. Both Deleuze and Heidegger share an unease with thinking of the organism as a self-defined being—as found, for instance, in the mechanist and vitalist traditions—but both also point in different ontological directions. For his part, Heidegger clarifies where his emphasis does and does not lie: "Thus the organism is neither 'a complex of instruments,' nor a union of organs, nor indeed a bundle of capacities. The term 'organism' therefore is no longer a name for this or that being at all, but rather designates a *particular and fundamental manner of being*" (342/235).

Prior to any strictly biological description of an organism, we therefore learn that organisms have a more definitive feature that gives them their "self-like character." Heidegger's analysis has demonstrated that organisms are much more than an internally additive structure: the organism does not amount to an equation of propulsion 'organ' + respiratory organ + digestive organ + metabolic organs + and so forth.[5] And yet the organism is a structural relation, albeit not of this type. The organization of a life form has its derivation in the essence of captivation. His account also says more than the possibly trivial statement that the organism is more than the sum of its parts. It depends on how one speaks of the "sum," or, as he tentatively writes, the "self-like" character of an organism. A self, or selfhood as such, is specifically reserved for human beings. Yet there is undeniably something that gives the organism the appearance of a 'self,' something that is "proper" to the organism and it alone. The capacity an organism has for being driven toward some activity offers a clue here. The capacity itself is something proper to itself, something that refers to the organism as a whole without entailing the qualification of self-consciousness, self-reflection, or any other self-relation. Heidegger describes capacity in terms of what is properly peculiar to that particular living being. "*Proper peculiarity* [*Eigen-tümlichkeit*] is a fundamental character of every capacity. This peculiarity *belongs to itself* and is absorbed [*eingenommen*] by itself. Proper peculiarity is not an isolated or particular property but rather a *specific manner of being*, namely a way of *being proper to oneself* [*Sich-zu-eigen-sein*]" (340/233).

This description of how an organism is its own proper 'self' approximates quite closely the special manner in which Dasein becomes its own authentic self. The English translators highlight the etymological link between the organism's "proper peculiarity [*Eigen-tümlichkeit*]" and Heidegger's description of Dasein's "authenticity [*Eigentlichkeit*]" in *Being and Time*. In both scenarios, it is a matter of being proper to oneself in one's own being. Just like the capacity precedes the organ, this manner of being precedes the organism. Michel Haar captures Heidegger's point quite well: "There is no life without this self-referential unity of 'belonging to—its own': even the most elementary living beings, like amoebas, sponges, protozoa, have 'property'—that which is its own." Haar continues by noting that "*Property* as such (*Eigentum, Eigen-tümlichkeit*), or the capacity of self-possession, is, in the essence of life, more original than the organism, for it is not one quality among others but a '*mode of being*'" (26–27). The ontological description of an organism's being proper to itself brings us back to the guiding formulation of the organism as defined by "a *particular and fundamental manner of being*." The being of an organism as properly peculiar precedes any physical or biological description and thereby captures the unity of an organism as such. It *is* itself in the unity of its instinctual drivenness. However, we are in the process of seeing that the animal's proper peculiarity is essentially different from Dasein's authenticity.

A better way to describe the peculiar manner of being that is proper to animals—namely, captivation—is "as a form of *behaviour* [Benehmen], as a form of *self-like behaviour*" (GA29/30, 345/237). With the introduction of this ethological dimension, Heidegger is now in a position to draw an important ontological distinction between animals, stones, and humans. The manners of being unfold according to how we understand this self-like behavior. Lest we be led astray, only animals behave. Such is their ontological fortune. Stones and humans do not—at least not in the manner presented here. A stone does not behave at all, while humans could be said to behave, sometime well, sometimes badly, though always in an 'open' fashion that Heidegger captures by the term "*comportment*":

> But *our* behaviour—in this proper sense—can only be described in this way because it is a *comportment* [Verhalten], because the specific manner of being which belongs to man is quite different and involves not behaviour but *comporting oneself toward*. . . . The *specific manner* in which man *is* we shall call *comportment* and the *specific manner* in which the animal *is* we shall call *behaviour* [Benehmen]. (346/237)

"*Our*" behavior is different from *their* animalistic behavior. We behave differently. Even more so, Heidegger affirms that behavior and comportment "are fundamentally different from one another." There is, in other words, an onto-ethological difference between animals and humans. Particular attention should be paid to Heidegger's original German (*Benehmen* vs. *Verhalten*), for, besides etymological differences, there is a possible source for confusion once we begin to take note of the common French translations for behavior, namely, "*comportement*." An obvious site for potential problems can be found in Merleau-Ponty's *The Structure of Behavior*, a translation of his original *La structure du comportement*. I would not want to gloss over this difference, where the French "comportement" is translated as "behavior," which is precisely the essential terminological distinction that is made in English between human comportment (*Verhalten*) and animal behaviour (*Benehmen*). I will have cause to readdress this in the chapter on Merleau-Ponty.

Leaving human comportment to one side, though always in view, we need to look more closely at animal behavior. Heidegger primarily highlights how behavior implies a certain self-retention, a mode of "*driven performing*" as opposed to the free "*doing and acting*" that human beings elicit. There are limits to behavior, but not just in the sense that one can behave this way and not that way. The limits are more fundamental in that an animal's behavior is limited with respect to its own 'self'-relation. The animal cannot get out of itself; it cannot, as Heidegger will claim elsewhere of human existence, transcend itself—neither temporally nor toward the world. But this is getting ahead of ourselves. With behavior, there are definite limitations in an animal's capacity to be itself. "Behaviour," Heidegger writes, "is precisely an *intrinsic retention* and *intrinsic absorption*" (347/238). The primary limit imposed is that the animal is absorbed in itself, that is, its behavior is characterized by its inability to free itself from its encompassing capacities. This limiting factor of behavior will become a defining feature of captivation as the essence of the animal:

> Behaviour as a manner of being in general is only possible on the basis of the animal's *absorption* in itself [Eingenommenheit *in sich*]. We shall describe *the specific way in which the animal remains with itself*—which has nothing to do with the selfhood of the human being comporting him- or herself as a person—this way in which the animal is absorbed in itself, and which makes possible behaviour of any and every kind, as *captivation*. (347/238–39)

Each characteristic that accumulates around the animal has been tied to this essential "structure" of captivation, which is the "*essential moment of animality* as such." Organs, capacity, drive, proper peculiarity, behavior, absorption: all tend toward the specific and fundamental manner of being animal. Heidegger is careful to distance captivation from a passing state that an animal may fall in and out of, like we're accustomed to saying about ourselves when we are momentarily captivated by a painting or person. Captivation, as essential, "is the inner possibility of animal being itself." The animal *is* captivated in advance "as a whole in its unity."

But, in describing the animal in this way, we are now led to ask how the limitation on the animal's being is curtailed by such captivated behavior. What is suggested by the animal's being "as a whole in its unity" in captivation? To what does this "as a whole" relate? Does the being of the animal—as revealed through its behavior—uncover something like world? A few examples may help clarify Heidegger's entrapment of animal life.

THREE BEES AND A LARK

The animal does not relate to world; rather, its behavior suggests a different relational structure: "Captivation is the condition of possibility for the fact that, in accordance with its essence, the animal *behaves within an environment* [Umgebung] *but never within a world* [Welt]" (GA29/30, 348/239). Within the context of defining the animal in terms of being proper to itself, being absorbed, having self-retention, and other such 'inward'-directed characteristics, this sentence comes off as slightly incongruous, particularly since it receives no further explanation. However, it does point the way toward understanding captivation as something more than a self-retentive structure. It implies, most important, that the animal is captivated by something other than itself, or, even better, that it is captivated by the other that it itself *is* as captivated. In his book *The Open*, Giorgio Agamben invokes the etymology of Heidegger's language surrounding the animal to help link captivation (*Benommenheit*), absorption (*Eingenommen*), and behavior (*Benehmen*).[6] Each term relates back to the root verb *nehmen*, to take. Heidegger intentionally plays on this etymology to situate the animal as being taken (absorbed, captivated, transfixed, benumbed), not only with itself, but in being taken by its surroundings as well. This language presents the possibility that animals, as defined by captivation, relate to something other than themselves, even if it is not world as such.

At this point, we find Heidegger in need of an explanation for captivation in terms of an animal's relationality. The animal sees, hears, walks, waits, and, in so doing, always relates to . . . something. What is related to is left elliptical and for the moment unanswered. But there is a relation nevertheless. The analysis therefore shifts somewhat, and is directed less toward the 'inner' organization of an organism and more so toward the 'outward' connection of an animal with its immediate surroundings. To be fair, there is really no such distinction between inner and outer in Heidegger's discourse, other than the tone of his language: he'll speak less in terms of 'absorption in' and more on the 'relation toward.' In this respect, his focus changes. "The task," Heidegger now writes, "is to see precisely *what kind of relationality* [Bezogenheit] lies in this behaviour; to see above all how the relationality of the animal's behaviour toward what it hears and what it reaches for is distinguished from human comportment toward things, which is also a relatedness of man to things" (GA29/30, 350/240). This task is of particular consequence because it is in the structure of relationality that one discovers the fundamental ontological difference between animals and humans. This difference, I would add, hinges on the ethological dimensions of behavior and comportment, and is thus an ethological difference with ontological repercussions. Up until this point, Heidegger really has yet to offer sufficient reason for claiming an essential difference between animals and humans. Remarks have been made, but with little to no support. One could still argue that both animals and humans are biological organisms, even if Heidegger finds the concept of organism problematic in its modern usage. Yet to do so, one clearly interprets animals and humans as comparable only insofar as one considers them as ontic entities. This said, however, there is still an underlying difference, an essential difference, and one that appears in the manner of being particular to the animal and human. This manner of being, as specific and fundamental, is discovered in animal behavior.

To better demonstrate his claims, Heidegger draws "concrete examples" from Uexküll's *Theoretical Biology* as well as Emanuel Radl's *Investigations on Animal Phototropism*. The primary example of animal behavior is an interesting choice: it is not that of a 'higher' order mammal, such as the behavior of chimpanzees or apes, or of dogs or rats, but instead it is that of the bee. In a way, Heidegger's choice of example should really have no bearing on his analysis, since he is concerned with the essence of animality as such, irrespective of the animal. The bee is neither too questionable, as would probably be the case had he used 'simpler' animals such as amoebas, protozoa, or even the glowworm that he describes earlier in terms of retinal vision. Nor is the bee too controversial, as might be the case had he used an example of chimpanzee or ape behavior, which, in more ways than one, demonstrates an affinity with human beings. This, it ought to be noted, will not stop Heidegger from claiming in a later lecture course that apes don't have hands, but only prehensile organs for grasping: "Apes, too, have organs that can grasp, but they do not have hands. The hand is infinitely different from all grasping organs—paws, claws, or fangs—different by an abyss of essence" (GA8, 16). Despite the biological proximity of the ape to humans, the essential difference is an infinite one, and it has everything to do with how each being relates to its environment. Bee or ape, it doesn't matter for Heidegger. The example of bees happened to be convenient (it was offered by his favorite biologist, Uexküll), and, not to mention, their behavior has often been a source of human wonder. Aristotle, to speak of favorites, has quite a bit to say regarding bees. In *Historia Animalium* he remarks on our inability to recognize how bees communicate with one another: "Each bee is followed on her return by three or four bees . . . how they do it has not yet been observed. . . . Their working methods and way of life show great complexity" (623b27). It was not until Karl von Frisch came along, some two thousand years later, that his studies of bee dances would begin to unlock the complexity of bee communication. Echoing such long-standing fascination, Heidegger may have put his finger on it best when he states concerning bees, "the situation is very simple—and yet thoroughly enigmatic" (GA29/30, 351/241). Our curiosity is piqued.

Heidegger presents three different scenarios that depict a bee's behavior, the first is an everyday, amateurish observation of the bee and its flying abilities, while the latter two are based on experimental research conducted on bees. The first example is one that we can all relate to since it requires only our ability to observe a bee's activities: a worker bee flies around a meadow from one clover blossom to another, all the while skipping over other flowers entirely. The bee remains "intent" and "consistent" with respect to its flower; *"blumenstet," "stetig," "Blumenstetigkeit,"* are some of the ways Heidegger qualifies such steady behavior. Uexküll reminds us that certain flowers are "significant" to the bee, while others are not. The bee enters into a duet with this flower as opposed to that one. To be more precise, however, it is not the flower as a whole, let alone the flower as such, but a certain scent that is indicative, as well as the flower's color, though the color, we are told, is apparently a little less relevant. Might one also say that it is the form of the flower that is attractive, along with its color? Perhaps, but we are only informed that it is the color that is visually distinctive. The anatomical structure of the bee's eye and its capacity for seeing only allow for "inferences" on our part, so at best we can only hazard a guess.[7] Once attracted to a given flower, the bee stops, "finds a drop of honey," sucks it up, and leaves for either another flower or for the hive. Two questions emerge from this simple scenario: does the bee recognize the absence of honey, such that it knows when to fly away? And second, if we admit that the bee recognizes honey as available in a given flower (for the bee does fly to *that* flower after all, and not another one), does this prove that the bee recognizes the honey *"as present"*?

These two questions are of particular note, for they get right to the heart of the matter. The issue is whether a bee is capable of detecting the absence or presence of something within its environment. Does the bee, in other words, *comport* itself toward the flower and honey, or does the bee *behave* and remain fixed in its relation? Before this can be answered, though, we discover a small slip of the pen, as Freud would say. We're given a simple sentence: *"Die Biene findet z.B. in der Kleeblüte ein Tröpfchen Honig"* (GA29/30, 351). While Heidegger focuses his attention on whether the honey is recognized as absent or present by the bee, it is easily glossed over that the flower itself does not contain a drop of honey at all. A flower contains nectar, that only later is converted, by bees, into honey. To say that the bee finds honey in its flower *may* be true, but only if we grant that the bee recognizes nectar as the potential source for honey that it will produce later. If this were the case, this certainly requires some degree of foresight, not to mention recognition of nectar *as* the potential for honey. I would not be tempted to call attention to this seemingly small oversight, which is no more than a scientific and natural inaccuracy, were it not for the fact that both this and Heidegger's next example describe the bee's relation to the honey in the flower stem. In the midst of the absence and presence of honey, one large slip is made: in calling nectar "honey," Heidegger is in fact saying that the bee relates to an absent presence! Honey is the absence related to in the bee's relation to nectar. So whether the bee relates physically to the nectar it is sucking up, which is Heidegger's emphasis, pales by comparison to this suggestion that the bee relates to what is not yet present in the nectar.

Does this undermine Heidegger's example? In the end, I don't think so, since the underlying questions remain unaltered: does the bee relate to something (e.g., the nectar, or flower, or hive, or sun) *as* present and/or *as* absent? Heidegger's slip does not affect the basis for this question. However, insofar as the foundation of his questioning rests on whether the honey is absent or present for the bee, it is more than ironic that honey, which is neither present nor absent in nectar *as such*, is taken as the measure of absence and presence. Either way, some degree of recognition of absence and presence seems to be already answered: the bee recognizes honey as an absent presence in nectar. The reason why I don't think this problem gains much momentum is that such behavior is easily found in any number of animal species, yet none of them speaks to the ontological foundation that Heidegger addresses. For example, the beaver or bird recognizes the potential of a tree or twig for the design of its home; the orangutan recognizes the potential of a stick to guage the depth of water it wishes to wade through. Such examples demonstrate tool use among animals, not to mention the possibility of culture (as learned behavior passed down through generations), on top of the recognition of a potential other use for something found in nature (a twig is something more than a twig: it is a house, an implement, a toy). From the standpoint of ontology, however, characteristics such as these matter little if there is no sense for how the living thing (bee, beaver, orangutan, human) relates toward the thing given. How open is the manner of access to the thing in question? Is the nectar present *as such*, that is, as a being at hand in its readiness to hand? Unless the nectar is given *as* nectar, which is admittedly still a questionable necessity at this point, the bee does not relate to the nectar. This type of relation is what is at stake in these examples.

The second example is taken from experimental research. With a bowl of honey placed before a bee, a scientist carefully makes an incision in the bee by cutting off its abdomen while it is actively sucking up the honey. Though Heidegger only notes the loss of the abdomen, we can also assume that this incision would disturb the biological receptor that lets the bee know that it is satiated. This experiment gets right to Heidegger's second question because, without its abdomen, the bee does not physically sense that it is full (its biofeedback system has been disrupted), and so it keeps sucking the honey. The answer to the question "Does the bee recognize the honey *as* present?" seems to be an unequivocal "no." The bee does not relate to the honey as present because the bee keeps sucking it up, without noticing the amount of honey it ingests. Heidegger explains that such an experiment demonstrates how "the bee is simply taken [*hingenommen*] by its food. This *being taken* is only possible where there is an *instinctual* 'toward...'" (352/242). The bee doesn't register the vast amount of honey present before it, and instead keeps instinctually sucking up the honey far beyond the amount it would naturally suck up or even need.

At first thought, this behavioral response would differ little from the action of a human. Take away my stomach and with it my sense for whether I'm full or not, and I too may eat myself silly. Sure I may recognize that I'm eating far more than normal, but based only on observation, how can we deny the same of the bee? Something more convincing is required for Heidegger's argument to work. Not only that, but one would have to think that the removal of the bee's abdomen is not simply the removal of a part from the whole, which would suggest a mechanist view, but that its removal affects the bee's being as a whole. Remove an arm or a leg, to say nothing of an internal organ like a stomach or the bee's abdomen, and you witness a disruption in the overall behaviour. On this point, Heidegger might have something to learn from Merleau-Ponty's phenomenology of the body. Consider, for instance, Merleau-Ponty's frequently cited examples of someone suffering from a phantom limb (PhP, 90–91/76). The case here is that someone has lost an arm or leg, through war or maybe for medical reasons, but continues to experience the limb as if it were still there as part of the body (e.g., a person may still 'feel pain' in the missing arm). It is worth noting that Merleau-Ponty also draws a comparison to the case of an insect that, having one leg removed, continues to act by substituting another leg for the missing one. The insect, like the amputee, "continues to belong to the same world" in its bodily being. Although Merleau-Ponty does not equate the insect with the human, they both continue to relate to their respective worlds despite the absence of the organ.

On one level, such a scenario surely asks us to question just how attuned we are physiologically to our surroundings. In the case of the bee, Heidegger's point is that the bee does not recognize the honey *as* present because it keeps sucking up the honey long after its abdomen has been removed (it senses no satiation). Uexküll, in another example, relates how rats devour their own legs if their nerve senses are taken away (TB, 145). Does this mean that a rat does not relate to its own leg *as* its leg, but as if it were like any other thing to be gnawed on? This experiment is conducted on only one lab rat, but what might happen if one observes the rat within its life as a pack animal, as Deleuze and Guattari are fond of highlighting? I wonder if this is really the point once key physiological receptors are removed. Such an example does not exclude human comportment, to be sure. Again, I question Heidegger's judgment on such an example since it removes a key organ (the abdomen) and then asks about the bee's capacity to relate to its food *as* present. Surely the absence of the organ affects the organism as a whole, even though the bee continues to inhabit the same environment as before. In part, Heidegger acknowledges such an issue, noting that the physiological process "is a matter of controversy" within the sciences, but, even still, it "is not the decisive issue for us here" (GA29/30, 353/242). Even if it isn't the decisive issue, the artificial nature of this study provides the basis for Heidegger's conclusions. He explains that the instinctual behavior of the bee to keep sucking displays a relation to the honey, but not a recognition of the honey as honey: "This is not to deny that something like a directedness toward scent and honey, *a relation toward* . . . , does belong to behaviour, but there is no recognitive self-directing toward these things. More precisely, there is *no apprehending* of honey *as* something present, but rather a peculiar captivation that is indeed related to the honey. The drive is captivated" (353–54/243). The final sentence is particularly revealing. The bee is reduced to the singular drive of sucking up honey, and it is the drive that is described as captivated, rather than the bee. This is because the bee *is* its driven behavior. The bee is unable to relate to the honey as such because it is always already captivated, in its essence, by its driven behavior. It does not register the presence or absence of honey because it is so intent on sucking up the honey that all else is neglected. But this doesn't yet explain why captivation necessarily precludes the all-important relation to things *as* such. This distinction is revealed through the third example.

In returning to a normal bee, abdomen intact, Heidegger draws attention to the fact that the bee must eventually stop sucking up the nectar and return to the hive. Does this mean that the bee's captivation ceases when it stops sucking? No! On the contrary: "The instinctual activity is simply redirected toward flying back to the bee hive" (354/243). The bee is always already captivated; it is not a matter of switching off one captivation and turning on another. Captivation, we will recall, is its fundamental manner of being. With this third example, then, the question is how the bee has "the *capacity for returning home.*" This suggestion that the bee is not at home in flying about the meadow is intriguing, for where might home otherwise be? This is all the more interesting if we take note of a reference Heidegger makes to Novalis, with which he commences his lecture course: "Philosophy is really homesickness [*Heimweh*], an urge [*Trieb*] to be at home everywhere" (7/5). There is a peculiar relation between the two in that neither the bee nor the philosopher are at home, and both share this urge or drive to be at home. Further, this concern with the home appeals to the very ontological status at stake in relating animals and humans—namely, their relation to the world. Heidegger writes: "Philosophy can only be such an urge if we who philosophize are *not* at home everywhere" (7–8/5). Particular emphasis should be paid to the connotation of "everywhere." In contrast to not being at home everywhere, which is the condition of "we who philosophize," Heidegger writes: "to be at home everywhere means to be at once and at all times within the whole. We name this '*within the whole*' and its character of wholeness the *world*" (8/5). To start his lecture course in this manner is particularly evocative, especially in consideration of his descriptions of the bee. The philosopher's position is one of *not* being at home. This trope of being lost, caught in an aporia, or not being at home, is common enough throughout the history of philosophy. One finds a reference to the philosopher as one who is lost and not at home throughout Plato's works, such as in *The Republic* and the *Protagoras*.[8] Descartes famously loses himself in a whirlpool of doubt only to find his way out again. Deleuze too suggests that philosophy and the philosopher are comparable with "the idiot" who "lacks the compass with which to make a circle" (DR, 170–71/130). Is the philosopher therefore like the bee as one who is not at home in the world? Does Heidegger's philosopher become-bee much like Deleuze's philosopher becomes-tick (LS, 158/133), as we shall later see?

There is a distinction to be made here. The bee is driven to be at home, by which Heidegger really means its hive. Why does he say that the bee is not at home in the meadow when this seems to be the most natural assumption? Though he offers no reasons, one would have to conclude that it is because the bee is not homesick in the sense of the world as a whole. To be at home everywhere means to be "not merely here or there, nor even simply in every place, in all places taken together one after the other" (GA29/30, 8/5). This description falls in line with what Heidegger has to say of the world: the bee is not at home in the meadow because it is not at home everywhere. But for Dasein to be at home, it is not simply the case, as it is for the poor homeward-bound bee, to return to a hive or house, or to be in the company of friends or family. If home is neither here nor there, where then is it? "This is where we are driven in our homesickness: to being as a whole" (8/5).

This much accords with Heidegger's analyses on animal being and human Dasein. The bee is captivated by its drive homeward. It is not presently at home, but on its way, en route. And yet how does Heidegger describe the philosopher's relation to the world? In a similar manner, namely, as en route, in a transition. Continuing from the previous quote: "Our very being is this restlessness [*Getriebenheit*]. We have somehow always already departed toward this whole, or better, we are always already on the way to it. But we are driven on [*angetrieben*], i.e., we are somehow simultaneously torn back by something, resting in a gravity [*Schwere*] that draws us downward. We are underway to this 'as a whole.' We ourselves are this underway, this transition" (8/6). If the philosopher, or Dasein, is under way toward the whole, driven on yet also torn back, does this mean that Dasein, like the bee, does not yet have a world? Are we to rethink what it means to 'have' a world? Even if Dasein's relation to home as a whole is ontologically different from the bee's relation to its home as hive, does Dasein's homesickness translate into a poverty in world? What, moreover, are we to make of the "contemporary city man, the ape of civilization," who has abolished homesickness? If one is not homesick, then one is an animal, such as a bee or an ape. But in being homesick, one is simultaneously driven toward and torn back from home. Dasein is driven, under way, a transition, a direction, headed toward home, but never at home in the world. Is there another dimension by which we ought to understand Dasein's dwelling as fundamentally '*unheimlich*,' where *unheimlich* denotes something beyond the uncanny? Is Dasein then an animal by another name, deficient in world and vexed to be at home in it?

Heidegger wouldn't think so, to be sure. Well, at least not as it concerns Dasein's being animal. While the bee is absorbed in its behavior, Dasein is gripped by its ontological conundrum of being-at-home in the world. This does not mean that Dasein is an animal, conceptually at any rate. However, it does appear to destabilize Dasein's relation to world. The bee, on the other hand, is never at home in the world *as* a whole, and where home is equivalent with hive, the bee is en route to a very physical location. What might happen if the physical locality of the bee's home is moved is the subject of Heidegger's third example.

The third experiment Heidegger reports is that of the bee's return flight home. Various studies have been conducted on how a bee returns to its hive and each offers a different perspective on bee behavior. In some instances, the hive is moved a few meters from its original spot. This experiment illustrates that the bee flies back to the original spot of the hive, but after becoming "suspicious" (though not "restless") with respect to its whereabouts, the bee eventually discovers the hive just a few meters removed. There is no mention here of whether the bee registers the absence of the hive, only that there is an "empty spot." Phenomenologically, one might wonder how the empty spot appears in the bee's visual field. The emptiness may be comparable to the absence of an expected thing against the Gestalt of the environment: the hive is missing from its context. Though wouldn't this suggest a scenario that Heidegger would not be inclined to extend to the bee, namely, some bare recognition of absence (not to mention an expectation or anticipation, i.e., some relation to the future)? How closely might this case of the missing hive approximate Sartre's celebrated example in *Being and Nothingness* of waiting for his friend Pierre in a Parisian café (9–10)? While Sartre sits and waits, Pierre's absence is pronounced against the background of the whole café. It is not that he is missing from either here or there; he is missing from the café as a whole. As much as one might like to think there is a similarity here, this isn't a sufficient analogy either, for, unlike Sartre, who does not discover Pierre's absence in a precise spot within the café (Pierre is missing from the café as a whole), the bee does discover the hive missing from a precise spot (it is missing from *that* spot where it is expected). I am inclined to think that the hive must nevertheless appear, as an "empty spot," against the background of the environment as a whole. Yet I am also inclined to think that there is a "flickering of nothingness" (10) by which one might imagine the dawning of absence, even though this is neither what Sartre intended by this phrase, nor what Heidegger would agree to by the term "empty spot."

The question here is how the bee managed to find its way back to the precise location of the hive's original spot. It is noted that the color of a hive is one feature, and so too is the scent that a bee emits. However, given that a bee's range can cover up to three or four kilometers, these features are not enough. As it turns out, a bee orients itself by means of a "landmark" (*Wegmarken*), though this mark is not of this land: it is the sun. The next study demonstrates this with the following case: since a bee normally returns to its hive a few minutes after having left it, the movement and temporal lapse of the sun were never considered a factor. However, if the bee is trapped in a dark box while it is away from the hive, and remains trapped for a significant amount of time, one observes that the bee will retrace its route back to where it thinks the hive is based on its relation to the sun. But if the sun has changed position in the sky by a 30 degree angle, the bee will be off the mark by a comparable 30 degree margin. In addition to the direction, the bee will also judge its distance back to the hive on the basis of the sun. Heidegger further notes that the bee "does not fly disorientedly or indiscriminately in any direction whatever," but is driven and stays the course back to where its home should be.

So what does this account of the sun tell us about bee behavior? Namely that the bee is captivated by the sun. It might be tempting to say that the bee has transferred its instinctual drive from the nectar to the sun, and then on to something else, but this seems awfully reductionistic, even if this sometimes seems to be what Heidegger means. "The bee is *simply given over* to the sun and to the period of its flight *without being able to grasp either of these as such*, without being able to reflect upon them as something thus grasped" (359/247). We could probably expect this answer from Heidegger at this point, but what comes as a surprise is the initial reason that he offers to account for the bee's driven behavior. He begins by noting the bee's self-absorption in going about its daily business: "Rather [the bee] is absorbed by a direction, is driven to produce this direction out of itself—without regard to the destination. The bee does not at all comport itself toward particular things, like the hive, the feeding place and so on." But now note the language that he uses in the next passage, as well as the final bewildering sentence: "But one might object that the bee does comport itself [*verhält sich*] toward the sun and it must therefore recognize the angle of the sun. It should be clear that here we become involved in insoluble difficulties [*unlösbare Schwierigkeiten*]" (359/246). These are seminal passages in Heidegger's lecture course, ones that reflect Heidegger's engagement with concrete experiments—"one of the most fruitful things that philosophy can learn from contemporary biology" (383/263)—but, as David Krell (126) has remarked, Heidegger not only pulls up short in the face of difficulty but also lets slip that bees *comport* themselves.

It is all the more intriguing and equally frustrating, therefore, that at this stage Heidegger states that "one could say," for example, that bees comport themselves toward the sun, but that if one were to say this, it would only present insoluble difficulties. Comportment, "the *specific manner* in which man *is*," is all of a sudden suggested with respect to the bee's relation to the sun, but it is not so much rescinded as named an aporia and swiftly passed over. What does Heidegger do next? He stays the course, just like a bee, driven onward with his analysis: "Nevertheless, we must not turn away and abandon the attempt to illuminate this peculiar and characteristic behavior" and so on. Further, and I'm tempted more and more to think that Heidegger himself has become all turned around in these analyses, he writes that "no advance is possible as long as we regard this behaviour as an isolated phenomenon in its own right." Instead, "we must ask therefore what we can learn about the general characterization of [animal] behaviour" (GA29/30, 359/246–47). Suddenly a single bee is not enough to draw a conclusion about all animal behavior. Whereas earlier what was said of one animal's essence held true for "*all* animals, *every* animal," the possibility of the bee's peculiar comportment toward the sun now makes this example more suspect. This isolated phenomenon was intended to offer clarification on the essence of animal behavior, but it is now being withdrawn or, more accurately, drowned out by a more general abstraction by speaking conceptually about "the animal itself." And, with this, the captivated bee is left behind in insoluble difficulties in place of a return to the animal as such. Quite literally, the bee is left in this mire, never to appear again.

Interestingly, Heidegger pulls a similar maneuver in his comments on "a lark" in his 1942–1943 *Parmenides* lecture course. Before resuming our interpretation is worth taking a brief look at this parallel. Both Michel Haar and Giorgio Agamben offer readings of Heidegger's lectures on Parmenides, particularly the final sections where Heidegger takes exception with Rilke's characterization of the animal's ability to see "the open" in his eighth Duino Elegy. But within Heidegger's reading there is a peculiar passage that offers a parallel to the one I just recounted with respect to the bees. Heidegger writes: "it would never be possible for a stone, no more than for an airplane, to elevate itself toward the sun in jubilation and to move like a 'lark,' which nevertheless does not see the open. What the lark 'sees,' and how it sees, and what it is we here call 'seeing' on the basis of our observation that the lark has eyes, these questions remain to be asked" (GA54, 160). But just as abruptly as with the bees, Heidegger changes topics when it comes time to say what the lark does see if it doesn't see the sun as such. So, once again, precisely when it becomes a task of accounting for the relation an animal has with, in this case, the sun, Heidegger draws back. Is it coincidental that in both scenarios it is the sun that instigates unanswerable suspicion? Might it be possible that the bee and the lark, to name just two animals, have an unparalleled access to the sun's illumination, to which we don't have access? Twice Heidegger notes that the animal does not "openly" see the sun as such, but when it comes time to offer an alternative account there is a lapse. I would not expect or want Heidegger to "speak for" the animals, to say what they see.[9] I would have expected, however, for him to say more than that these animals do not have access to the open, without qualification. Why don't the lark or bee see the sun as such when they presumably have the eyes to do so? It requires more than just eyes to relate toward the sun, but what does Heidegger have in mind then? What inhibits these animals, to speak of just the two, even though we know that they represent countless more, from accessing a more fundamental relation to their surrounding world?

ANIMAL MORPHOLOGY

Though it at times seems that Heidegger gets all turned around in his analyses of concrete biological examples, he is not unaware of the difficulties in which he has immersed himself. He admits that the behavior of animals is "an enigma" because it "repeatedly forces us to address the question: What then is behavior related to and what is the nature of this relation?" (GA29/30, 367/252). Heidegger's entire reflection on animal life has been heading toward this very enigma, that of animal behavior, which appears both familiar and distant to us. In a way, the discussion comes down to a language of circles and spheres, with the animal at its center (276/187). The animal, as Uexküll warned, is obliviously ensconced in its spherical environment, leaving Heidegger to circle continuously around in it in the hope of gaining greater insight (267/180). But when the dizzying ambiguity becomes too vertiginous, Heidegger tends to pull back, and, just like an intrepid explorer who marks a particular crossing with a flag, he leaves a mark for future retracing: "insoluble difficulty," "enigmatic," "impenetrable," "these questions remain to be asked." Heidegger nevertheless continues to circle the animal with an eye toward penetrating its sphere. What do animals relate to? And how is the ontology of animality revealed through this phenomenon of behavior?

Various examples suggest that animals are captivated by things in their environments: the sun, a flower, an animal, a rock. The animal relates to other entities, but not necessarily to entities as such. This captivated behavior provides a glance at how something can be open to the animal (e.g., the bee has a relation to the sun, the lizard to a rock), but how that very same thing is withheld from the animal too (e.g., the bee does not perceive the sun as such). The thing is withheld, but, as Heidegger clarifies, it is not closed off from the animal either. There *is* a relation, albeit a peculiar one: "The *captivation* of the animal therefore signifies, in the first place, essentially *having every apprehending of something as something withheld from it*. And furthermore: in having this withheld from it, the animal is precisely *taken by things*" (360/247). If we recall the case of the bee for a moment, a bee has access to the sun as something not accessible as such. The ontological significance of the sun withdraws as the bee relates to it. The problem for the animal lies in its inability to let things be—it cannot let the sun repose in its being. Nor, therefore, can it disclose the sun's being as being closed off. Rather, in its driven behavior, an animal is captivated by something from within its own transfixed state of being. According to Heidegger, every animal lacks the capacity to relate to things in their being and are therefore instead caught, "taken," or "suspended." This ambiguous relational state is probably best characterized by what Heidegger calls a little "leeway [*Spielraum*]" that an animal has in its access to things. There is a little give-and-take in what is open to the animal, but fundamentally the *"animal as such does not stand within a manifestness of beings"* (361/248).[10] "Leeway" captures the animal's captivation quite well: the animal is suspended between itself and its environment, it has and does not have world, it is open to things that are nevertheless fundamentally withheld, it offers a little room to play within but without the incursion of anything too grave. It is captivated, but not spellbound.

Again, much of this has to do with how Heidegger describes our own relation to the animal and how we might penetrate its inner essence. The language is not entirely innocent, for, in a way, the issue pivots on how we can "transpose" ourselves into the behavioral environment of animals and "penetrate" their inner core. There is almost an obscene violence to it, were not for the fact that Heidegger neither punctures nor wounds, but circles closer and closer around the animal's sphere. In this story, there are actually different kinds of circles and spheres, above all the philosopher's circle (*ein Kreis*) and the animal's circle (*ein Ring*). Just as Uexküll hypothesized that all animals live within a "soap bubble," Heidegger too relies on a spherical model to capture the overall structure of an animal's captivation with its environment. By the end of his analyses, Heidegger admits that Uexküll's *Umwelt* is really just a pale precursor to his own concept of "the disinhibiting ring" (383/263), but Agamben is probably more accurate when he suggests that Heidegger didn't fully recognize the extent to which his own terminology was influenced by Uexküll (51). Given how much Uexküll figures into the background of these lectures on animal life—he is there from the beginning (284/192) to the end (384/264)—it would be surprising if Heidegger did not owe his model at least in part to the whimsical *Seifenblase* Uexküll blew around the animal. But whereas Uexküll was blowing bubbles, Heidegger finds the rings far more secure in their makeup. The differences between their respective descriptions are in fact quite revealing. Uexküll harbored no intention of popping the bubbles that surround animals, for he was more content with observing their lives through the opaque transparency that separates each animal's experiential life. At best, there is always a metaphorical film or residue that separates our access to their environments.

With Heidegger, on the other hand, we encounter no disillusion in his wish to penetrate through the ring to get to the essential core of animality, even if he also contends that such a breakthrough may never be achieved. His language is suggestive when it comes to first describing the ring: the animal's drivenness "holds and drives the animal within a *ring* which it cannot escape [*einem* Ring, *über den es nicht hinausspringt*] and within which something is open for the animal" (GA29/30, 363/249). It is as though animals are imprisoned within their own being, from out of which they can never leap or spring free. Yet this ring is also open, so the ring does not enclose the animal solely in itself (it is not a solitary confinement). In another colorful description, Heidegger writes that the ring "is not like a rigid armour plate [*kein fester Panzer*] fitted around the animal" (377/259), nor is it "as if some kind of barrier were erected in front of the animal" (373/257). To be sure, the ring is not a physical appendage or protective sphere in which the animal finds shelter. Let us not be fooled; he is not speaking about the exoskeletons of armadillos, turtles, lobsters, or snails. He may have written about snails in his earlier 1925 lecture course, *History of the Concept of Time*, but he is not doing so here. In fact, Heidegger has changed his tone significantly in these four years. For instance, observe these earlier comments:

> The snail is not at the outset only in its shell and not yet in the world, a world described as standing over and against it, an opposition which it broaches by first crawling out. It crawls out only insofar as its being is already to be in a world. It does not first add a world to itself by touching. Rather, it touches because its being means nothing other than to be in a world. (GA20, 223–24/166)

There are a couple of things to note. The framework of Heidegger's thought has not changed: the relational structure remains unaltered, for it is still a question of how to frame this act of "being-in" as one of relating to the world. The snail, just as with Dasein, is not an entity set over and against the world. The snail's being, as John van Buren has remarked, is already enmeshed in its being-in-the-world.[11] However, Heidegger has reconsidered what the snail is 'in' by the late 1920s. Being-in-the-world is no longer applicable to the snail, whether it is in or out of its shell.

It is possible that Heidegger reformulates his view of the snail's world in response to some remarks made by Max Scheler in *Man's Place in Nature*. Observe the similarity between the two, in Scheler's words: "The animal has no 'object.' It lives, as it were, ecstatically immersed in its environment [*Umwelt*] which it carries along as a snail carries its shell. It cannot transform the environment into an object." Scheler continues by comparing the predicament of the snail to human behavior: "It cannot perform the peculiar act of detachment and distance by which man transforms an 'environment' into the 'world,' or into a symbol of the world" (34/39). Scheler captures the peculiarities of animals and humans in two simple diagrams, each representing either the closure of animal behavior—T (*Tier*/animal) $\leftrightarrows$ U (*Umwelt*)—and the openness of human comportment—M (*Mensch*/human) $\leftrightarrows$ W (world) →→. Despite the obvious parallel with Heidegger, the differences between Scheler and Heidegger are just as pronounced: while Scheler claims that the animal lives "ecstatically" in its environment (a claim that Heidegger does not share, as we shall see), he also claims that the animal cannot distance itself from its environment (a claim that Heidegger reciprocates). Further, even though Scheler emphasizes the "world-openness" of human behavior, he uses the same term for animal behavior (*Verhalten*), the very term that Heidegger will use to emphasize human comportment. But above all, Scheler's descriptions of humans as spiritual beings are antithetic to Heidegger's problems with metaphysical humanism and philosophical anthropology. Scheler thus provides an interesting contrast between Heidegger's earlier remarks on the snail and the 1929–1930 course, where the snail, along with all other animals, is placed firmly within a ring wherein it both has and does not have a world.

The possibility of the snail's "touching" is also questioned in the later lectures. The stone does not touch anything "in the stronger sense of the word," and even the lizard's sense of touching the stone is not the same as what is meant by a human's sense of touch (GA29/30, 290/196). If the snail's touch is determined by its being as being-in-the-world, one can bet that once the snail's being is reevaluated, as it is in these later lectures, so too is its sense of touch. Interestingly, Heidegger broaches this topic of animals and touch in *Being and Time*, particularly in its relation to time, but leaves it as a problem (GA2, 346/396). A further ontological clarification of animal life will wait to be amended until these later 1929–1930 lectures. The other point to note is that the snail's shell is not what Heidegger means by "the ring" that encircles it. The ring is co-constitutive of an animal's being and is like no present-at-hand substance that one could observe or, for that matter, touch.

These remarks on the snail are indicative of Heidegger's earlier interest in the worldhood of animals. Not only are snails described as being "in the world," but many other animals are too. In what may be an indirect reference to Uexküll, Heidegger, in a 1926 lecture in Marburg, comments on how even jellyfish already have a world. The world, in a certain sense, is uncovered and disclosed to the jellyfish.[12] Although unacknowledged, the reference may very well allude to Heidegger's reading of Uexküll's early work on jellyfish in *The Environment and Innerworld of Animals*, where one reads how "the organism is like a magic world, closed off [*verschlossen*] to all effects of the external world, opening [*öffnet*] only to the right key. If no lock is present, no key can be found" (UI, 71/230). For Uexküll, the animal is like a lock or answer, opening only to the right key or question. In the right structural relation, the animal certainly discloses itself as being of this environment. A further indication of this possibility, and one that only underscores Heidegger's turn away from the notion that animals might have a world, is a similar comment that one finds in his 1925 Kassel lecture, "Wilhelm Dilthey's Research and the Struggle for a Historical Worldview." In this lecture, Heidegger writes the following, which deserves full citation:

> Life is that kind of reality which is in a world and indeed in such a way that it has a world. Every living creature has its environment [*Umwelt*] not as something extant next to it but as something that is there for it as disclosed, uncovered. For a primitive animal, the world can be very simple. But life and its world are never two things side by side; rather, life 'has' its world. Even in biology this kind of knowledge is slowly beginning to make headway. People are now reflecting on the fundamental

structure of the animal. But we miss the essential thing here if we don't see that the animal has a world [Welt]. In the same way, we too are always in a world in such a way that it is disclosed for us. (SEE, 163)

What makes Heidegger's later retraction so curious is his emphasis on what is claimed to be the "essential thing": the main issue is the world, and always has been, but between 1925–1926 and 1929–1930 the essential difference has changed. In 1925, the essential thing is that animals have a world. By 1929, however, the essential thing is that animals are poor-in-world due to their being captivated by their surroundings. How can what is "essential" change? This is more than a dilemma; it is an outright enigma. From the two passages just cited, as well as the one on the snail, it appears as though Heidegger is loosely equivocating between *Umwelt* and *Welt*. One passage states that the *Umwelt* is uncovered and disclosed, while the other passage claims that a *Welt* is uncovered and disclosed. This discrepancy will be dealt with more smoothly by the time of *Being and Time*, and fully clarified by the 1929–1930 course. This fluctuation between having and not having a world on the part of animals comes down to Heidegger's descriptions of encircling rings, which also suggests a closer and more explicit dealing with Uexküll who remained unnamed in the earlier writings, despite Heidegger's usage of the concept *Umwelt*.[13]

So what exactly is this ring that "encircles [*umringt*]" every animal? Far from being "a kind of encapsulation [*Einkapselung*]" (GA29/30, 370/255), the ring is the structural totality that the organism *is* in its relation toward things in its surrounding environment. The connotation of this organic totality as a circular ring and as a "sphere [*Umkreis*]" unfortunately lends itself to the false interpretation that each animal is in some way circled by some invisible field radiating around the animal at its center. While this image may be helpful initially, the spherical metaphor suggests that everything within a given spatial arrangement is accessible to the animal, which isn't true. It isn't a spatial model at all, but more fundamentally a relational structure of the being of the animal. The spherical connotation, therefore, may ultimately prove to be more misleading in the case of Uexküll's soap bubble, but Heidegger's language may assist in this clarification. If not, we will have to wait until Deleuze and Guattari's "lines of flight" and "destratifications" arrive on the scene to puncture this model of circles and spheres. What Heidegger instead intends by an animal's self-encirclement is a manner of being by which every animal is capable of relating to those beings that "disinhibit [*enthemmt*]" its behavior, that is, " '*affects*' or initiates the capability in some way" (369/254). The double negative of "disinhibition" is deliberate:

animals are first inhibited in their relations such that only select beings may penetrate and *dis*inhibit their behavior. This disinhibition provides the opportunity for those select beings to become open to the animal.

Uexküll's example of the tick waiting for a passing mammal is a helpful one in this case. If we recall, the tick remains suspended on a branch. Many things might pass through a spatial proximity to the tick: rain, birds, heat, minerals, sunrays, and so on. However, the tick is unfazed by any of this, for none of these things belong 'within' the tick's ring. This condition lasts until at long last a dog or a human, for example, "disinhibits" the tick's behavior and "opens" the tick to the mammal. A significant relation is established from which we might understand an animal's self-encirclement. Aside from those beings that are capable of initiating each animal's specific disinhibitions, Heidegger is quite clear that "[n]othing else can ever penetrate the ring around the animal" (369/254).

Those who claim that Heidegger has nothing to say regarding the body would benefit by consulting his descriptions of animal life, for here one gains a perspective on the body due to its connection with the ring. The encircling ring (*Umring*), as Heidegger calls it, is a primordial feature of the essence of animality, and that "belongs to the innermost organization of the animal and its fundamental morphological structure" (371/255). It is worth pausing for a moment on the repercussions of this "disinhibiting ring" by which animals are simultaneously open to and closed off from an environment, by which there is both a relation and nonrelation between the animal and its environment. This concept of a disinhibiting ring becomes important not merely in terms of how it defines an animal's self-relationship and relation to other things, but also in how Heidegger seeks to express a different understanding of the animal as a corporeal organism. Drawing from his readings in biology, Heidegger notes how the theory of a disinhibiting ring creates a completely different conceptualization of the animal as an organism within an environment. Rather than believing that the animal actively encircles an environment around itself as it goes through life—which he likens to Darwin's theory of adaptation—Heidegger draws attention to the idea that this encircling is always already a part of the animal as a kind of pre-delineation *in and of* the very movement of life itself:

> Every animal surrounds itself with such an encircling ring, but it does not do so subsequently, as if the animal initially lived or ever could live without this encircling ring altogether, as if this encircling ring somehow grew up around the animal only at a later stage. On the contrary, every living being, however rudimentary it might appear to be, is surrounded in every mo-

ment of its life by such an encircling ring of possible disinhibition. (374/257)

Heidegger's note that this encircling is due to the animal's morphology further confirms that this ring around the animal does not solely belong as an environmental construction that the animal produces, but as an inherent structure rooted within the morphological development of the animal itself.

This point has tremendous implications when applied to conceiving the organic nature of the animal. What Heidegger argues, drawing on the investigations of Buytendijk and, more specifically, Uexküll, is that every animal's morphology already encompasses the environment within it. By finding the animal's behavior to be a capacity for relating to a limited environment, Heidegger makes the startling claim that an animal's life is not composed of a specifically confined environment becoming accessible to it, and to which it adapts, but rather that the animal comes to be itself, and have a self-encircled unity, in its being captivated with the environment. What is striking in this is how Heidegger erases the surface of the animal body as a false corporeal limit, and instead redraws the unity of the animal as one that already encompasses the environment within the morphology of the animal body. The border of the body of course does not become dispensable, but it is no longer seen as the primary limit that coheres the individual organism into a self-sufficient unity.

He first makes this claim in contrast to the claims of Buytendijk, whom he cites as writing the following: " 'Thus it is clear that in the animal world as a whole the way in which the animal is bound to its environment is almost as intimate as the unity of the body itself' " (375–76/258). There is a unity, Heidegger wants to claim, only it should not be limited to the physical body. Instead,

> if we consider the organism as the morphological unity of the [animal] body, which is what usually happens and especially today, then we have still failed to grasp the decisive structure of the organism. . . . Against this we must say that the way in which the animal is bound to its environment is not merely almost as intimate, or even as intimate, as the unity of the body but rather that the unity of the animal's body is grounded as a unified animal body precisely in the *unity of captivation*. (375–76/258)

What is addressed in this critique is where and how Heidegger seeks to question the totality of the animal's unity, and the conclusion toward which he heads is one that will ask us to rethink how the life of the animal is framed not by a corporeal unity that moves through space but rather how the environment is already encapsulated within the animal's morphology and behavior. This is the essence of the relation. Moreover, insofar as the body is unified as a whole in terms of its captivation, the morphological structure is itself grounded on the more primordial ontological structure of being encircled. "The *organization of the organism*," Heidegger writes, "does not consist in its morphological, physiological configuration, in the formation and regulation of its forces, but first and foremost precisely in the *fundamental capability of self-encirclement*" (375/258).

Here Heidegger most firmly aligns himself with Uexküll's discoveries in animal biology, over against those of both Buytendijk and Driesch. While Heidegger certainly believes that Buytendijk and Driesch offer positive contributions to contemporary biology, he also feels that they still do not fully appreciate the relational dynamic with the environment. Just as Buytendijk paid too much attention to the surface of the body as the locus of the animal's unity, so too does Heidegger find Driesch focusing too much on the body in his holistic descriptions of animal morphology. Driesch, Heidegger believes, is right in his conclusions that an organism exhibits its full and unified presence at every stage of its life (an observation that we already find concerning Dasein in *Being and Time* and earlier), but that Driesch succumbs to grasping the wholeness of the organism at the level of the body only. Accordingly, Heidegger notes that "we can see that the organism is certainly grasped as a whole here, yet grasped in such a way that the animal's relation to the environment has not been included in the fundamental structure of the organism. The totality of the organism coincides as it were with the external surface of the animal's body" (382/262). In the end, Driesch fails to consider properly the animal's ontological unity.

Heidegger finds more support in the findings of Uexküll, whom he feels offers an abundance of observations and descriptions that lend themselves to the philosophical argument that he himself is putting forward. Most important, Uexküll stresses the "*relational structure between the animal and its environment*" and "[i]n this connection the totality of the organism would not merely consist in the corporeal totality of the animal, but rather this corporeal totality could itself only be understood on the basis of that original totality which is circumscribed by what we have called the *disinhibiting ring*" (383/263). If the animal's totality is not confined to the body, then Heidegger discovers a new relational understanding between the animal and its environment, one in which the animal's essence can only be understood in terms of how the environment is always already there within the morphological structure of the animal itself—namely, in captivation. In speaking of animal life, Heidegger discovers not so much a firm delimitation of the body as an appendage within a particular environment, but instead perceives the animal as a being that already encompasses the environment within it, and, in so doing, prepares the way for reenvisioning the individuality of every particular life form. In whatever way one might wish to think of a unity or totality of an organism, one can do so only from the fundamental basis of captivation. The unity of captivation is, Heidegger writes, "a structural totality," which, moreover, "is the prior basis upon which any concrete biological question can first come to rest" (377/260).

That captivation is the basis for understanding the totality of the animal, particularly in its relational structure with an environment, is another reason for reconsidering even a physiological comparison between animals and humans. If, for Heidegger, animals and humans differ in terms of their behavioral relations, which is an ontological distinction, then this also necessitates a rethinking of their bodily kinship. In the 1946 "Letter on 'Humanism,'" Heidegger makes this very point: "The human body is something essentially other than an animal organism" (GA9, 155–56/247). It is a remarkable point, and one that may not sit well with many. Even from a biophysiological standpoint, Heidegger claims that humans are not the same organic being as the rest of animal life. Humans have eyes, legs, lungs, and so on, that compose an evolutionary link with animal life. However, there is an *essential* difference that is more primordial than physiological similarities. As we saw earlier, the organs are themselves grounded in the overall capacity an organism might have. Only now do we see the full repercussion of such a thought: the human body is essentially different from that of an animal's body, which is why it is not out of character for Heidegger to suggest that apes do not have "hands." This can be argued because of how animals are captivated by their environment and how humans comport themselves to their worlds. The unity or totality of a being is determined on the basis of this relation, and so too is our understanding of the body. Whereas a human body is understood from the basis of its essence as "ek-sisting," the animal body is intimately linked with its encircling ring.

A SHOCKING WEALTH

Such is the primordial nature of the ring. But we are still left wondering how, within the contours of this structural totality, something disinhibits the animal. An animal's relation to the world is not limited to its physical body, and it can be affected or stimulated by significant beings in its environment. But just because some things are capable of penetrating through an animal's ring, it does not follow that animals are open to these things as such. On the one hand, animal behavior is "eliminative, i.e., it is certainly a relating to . . . but it is so in such a way that beings can never, and essentially never, manifest themselves as beings" (GA29/30, 368/253). A mammal may be 'open' to the sensory organs of a tick, however it is 'open' only in such a way that the mammal "constantly withdraws" (370/254) because the tick's behavior "involves *no letting-be* of beings as such" (368/253). This applies for all animals. But why are animals unable to simply let things be? Why do things withdraw even from within the opening allowed for by the animal's self-encirclement? Though Heidegger will not be quite so explicit until his "Letter on 'Humanism,'" the answer to this is because animals lack language. In lacking language, an access to things "as such" will always be unattainable. Later in his 1929–1930 course, Heidegger remarks "[i]t is this quite elementary '*as*' which—and we can put it quite simply—is refused to the animal" (416/287). In being denied this linguistic copula, by which one can understand "something as something," things will never be manifest to animals in their being. The lizard, for instance, though open to the rock on which it lies, nevertheless cannot inquire into the "mineralogical constitution" of the rock (291/197). Nor can the lizard "ask questions of astrophysics" about the sun under which it basks. For this to occur, the animal must at the very least be in the possession of language, not to mention conversant in difficult scientific matters. I shall return to the issue of language shortly, specifically in its relation to transcendence and time.

Despite this eliminative characteristic of behavior, the animal's openness demonstrates on the other hand "a kind of wealth." The connotation of wealth lies in stark contrast to the animal's poverty, which is identifiably the source for ambiguity in describing the state of an animal's openness. After his characterization of animal life, Heidegger adds the caveat that animals are not thereby "inferior" to Dasein. Rather, "life is a domain which possesses a wealth of openness with which the human world may have nothing to compare" (371–72/255). In terms of Heidegger's thesis on the animal's poverty in world, Matthew Calarco has noted rightly that it implies no value judgment. Poverty, we are meant to believe, is not a comparison to Dasein's wealth in world, but the poverty of the animal in its own essential terms. But in the case of the animal's wealth of openness, there is the allusion, however vague it may be, to a wealth beyond compare. They have a wealth of openness that we simply cannot relate to, such as in how the bee or lark relates to the sun, how a dog leaps up a flight of stairs, how a bird makes its nest, or how a cat chases after a mouse (390/269). In each case the animal has access to beings in a mode of being-open, and it does so in a manner toward which we cannot relate. This is similar to the claim made by Thomas Nagel in his celebrated essay, "What is it like to be a bat?" We have nothing—in short, no experience—that allows us to compare ourselves to these different modes of access: they have a wealth that we cannot approximate. This is the very point behind Uexküll's *Umwelt* theory—namely, that discretion is required in observing the lives of animals because every animal inhabits a different environment than another. Heidegger's retort to such a claim is that this still does not mean that animals have access to things *as such*. But why does Heidegger qualify it as a "wealth"? And why does he do so particularly when the dominant metaphor in his analysis is that of the animal's "poverty"? Does this disrupt any possibility of a "comparative analysis"?

There are several possible responses, but each one equally results in more questions. One response has to do with the animal's disinhibition: the kind of openness an animal has to its environment is due to how it is disinhibited by things. In a negative manner, we at least gain a better understanding of how animals are "poor in world": "Accordingly we do not at all find a simultaneous having and not-having of world, but rather a *not-having of world in the having of openness for whatever disinhibits*" (392/270). Perhaps, then, whatever wealth animal life might have is still confined within the more general account of the animal's poverty, a poverty that is now understood to rule out any "having" of world. In the final analysis, which Heidegger admits is not the final word, animals do not *have* world. Instead they only have an openness for certain things that stimulate or initiate their behavior. If this is the case, how an animal is disinhibited has everything to do with its being-open.

As another possibility, the wealth of animals is therefore essentially linked to how they are affected by other things. In a curious passage, the captivated animal is "exposed" to something other than itself: "that which disinhibits, with all the various forms of disinhibition it entails, brings an *essential disruption* [wesenhafte Erschütterung] into the essence of the animal" (396/273). Unfortunately Heidegger does not describe this "essential disruption" any further. It is an evocative description, however, particularly when one considers the implications of what might be meant by *Erschütterung*. Translated as "disruption," it also carries the connotations of a "shock" (which is close to captivated or stunned), "distress," "tremor," and even "shattering."[14] Either way one looks at it, *Erschütterung* is an interesting way to depict what occurs in the instance of an animal's becoming disinhibited. The animal's openness, limited though it may be, essentially disrupts or shatters the very essence of the animal. More correctly, it is not so much a shattering of the essence of captivation—for this essence still holds as "*one* structural moment"—but that there is a fragmentation within the essence itself. One can only imagine what this disruption might be, this other that penetrates into the animal's essence. Is it the shattering of the ring that circles the animal, sundering any totality or closure this ring may have held? Could it be the gap or abyss, as an essential difference, that separates animal life from human Dasein? It may very well be that we need to wait for Deleuze and Guattari to outline more clearly the "shattering" of this ring around organisms, such as when they stress how an organic "stratum" shatters and fragments the organism's unity into gradations (ATP, 67/50–51). In the case of Heidegger, I wonder if it isn't the extensive range of human penetrability that in the end disrupts the very essence of animality that he sought to uncover.

In the final pages of the 1929–1930 lecture course, many different themes begin to converge around the function of the 'as' that the animal has been denied and is the source of the abyssal gap between beings and being, and animals and Dasein. The term "projection" is raised in order to speak of Dasein's "originary and properly unique kind of action" (527/363), but it is not so much Heidegger's use of projection that interests me here. Rather, it is his description of the 'as' structure and how this might relate to our reading of animal life.

> Yet—as the forming of the distinction between possible and actual in its making-possible, and as irruption [*Einbruch*] into the distinction between being and beings, or more precisely as the irrupting of this 'between'—this projection is also that *relating* in which the 'as' springs forth [*entspringt*]. For the 'as' expresses the fact that beings in general have become manifest in their being, that that distinction has occurred. The 'as' designates the structural moment of that originarily *irruptive* 'between.' (530–31/365)

The language Heidegger employs to speak of how projection "irrupts" into the 'between' is of particular note. We can recall how the animal could not figuratively leap out of or escape (*hinausspringt*) from its being, captivated as it is within its encircling ring. What is required is the 'as' that, we now discover, is itself the spring in expressing the ontological distinction. The 'as' of language is the spring that makes the connection 'between' being and beings. This spring is, moreover, a shocking one, an irruptive "structural moment" within the very being of Dasein. Indeed, the irruptions become so momentous that four irruptions and one eruption occur in just two sentences. This passage also indicates that Dasein, unlike the animal, knows how to make an 'exit':

> In the occurrence of projection world is formed, i.e., in projecting something erupts [*bricht . . . aus*] and irrupts [*bricht auf*] toward possibilities, thereby irrupting into what is actual as such, so as to experience itself as having irrupted as an actual being in the

midst of what can now be manifest as beings. It is a being of a properly primordial kind, which has irrupted to that way of being which we call *Da-sein*, and to that being which we say *exists*, i.e., ex-sists, is an exiting [*ein Heraustreten*] from itself in the essence of its being, yet without abandoning itself. (531/365)

I raise these passages from Heidegger's final pages because I believe they point toward a more fundamental issue than the distinction of language and the 'as' structure. Dasein is able to step out of itself in a way that animals cannot; animals remain transfixed in their manner of being such that no escape is possible. And even though Heidegger toys with the idea of an "essential disruption" within the essence of animals, this disruption is not the irruption of which he speaks here. Perhaps the vibration experienced in the essence of animality is that of one felt by the neighboring irruptions in the essence of Dasein. Dasein is the earthquake, the volcano, the "storm sweeping over the planet" (10/7), while animals may be the aftershocks registered a little while later, a little more feint, a little more ambiguous to read. The more fundamental issue is one that I believe Heidegger barely registers himself, at least not formally. By this I'm referring to time.

A FINE LINE IN THE RUPTURE OF TIME

Time is a surprising omission, if one can call it that. Considering its role in the ontological discernment of Dasein in *Being and Time*, as well as that the first third of the 1929–1930 lecture course is devoted to an analysis of boredom (*Langeweile*), it is notable that Heidegger rarely speaks of a temporal dimension of animals. Yet time is nevertheless lying quietly beneath this entire discussion. While I cannot enter into a lengthy analysis of animals and time at this point, nor of Heidegger's lectures on boredom, allow me a few reflections on what is said and not said on this topic.

With animals, the subject of time enters surreptitiously, though no less importantly. In fact, I'll argue that time is the very basis on which every other distinction between humans and animals comes to rest. Whether we consider Dasein's irruption of being versus the animal's disruption, comportment versus captivated behavior, acting and doing versus driven performing, language versus no language, Heidegger's distinctions implicitly draw on the temporal dimension of being. Our descent toward time is first revealed by how language owes its character to the transcendent nature of being. "There is language," Heidegger writes, "only in the case of a being that by its essence *transcends*" (GA29/30, 447/308). We have already witnessed that animals do not have language insofar as they cannot relate to other things as such. Shortly after stating in the "Letter on 'Humanism'" that animals lack language, Heidegger resumes his account of language: "In its essence, language is not the utterance of an organism; nor is it the expression of a living thing. Nor can it ever be thought in an essentially correct way in terms of its symbolic character, perhaps not even in terms of the character of signification. Language is the clearing-concealing advent of being itself" (GA9, 158/248–49).

Insofar as animals are not in the possession of language, they are 'nontranscendent' beings. Although Heidegger would reject the comparative designation of humans as the 'transcendent animal,' he does speak of the transcendence of Dasein, and there is no evidence that he would find exception to the nomination of animals as 'nontranscendent.' We have already observed how the animal remains captivated within its ring, unable to leap free and escape. Animals are trapped within their own being, unlike Dasein that can and even must step out from itself. The concept of transcendence is not overtly present in the 1929–1930 lecture course, but its import is at work throughout. The essence of transcendence is what allows for the possibility of a relation to.... An indication of the priority of transcendence can be found throughout Heidegger's writings of the late 1920s, such as in *The Metaphysical Foundations of Logic* and in "On the Essence of Ground," where he defines transcendence as a fundamental state of "surpassing [*Überstieg*]" as a whole. Transcendence can be summarized in many ways: as the "ground of the ontological difference" (GA9, 31/106), "as the *fundamental constitution of [Dasein], one that occurs prior to all comportment*" (33/108), that it "constitutes selfhood" (34/108), "is *world-forming*" (55/123), and is, in a word, "the being-in-the-world of Dasein" (59/125). In an important sense, however, it all comes down to the same: that transcendence is the fundamental constitution of Dasein's being-in-the-world wherein world "happens [*geschieht*]" through Dasein itself. And this, as Heidegger will later assert, "is rooted in the *essence* of time, i.e., in its ecstatic-horizonal constitution" (62/128).

The ontological repercussion is that Dasein is always already beyond itself insofar as it is "the *temporal entity simply as such*" (GA24, 383/271). Not only is Dasein ecstatic as the temporal entity, but "temporality [itself] is *outside itself*" (377/267). What I wish to emphasize is that Dasein is characterized as always "beyond itself," "outside itself," "stepping out of itself," "transcending itself," and that Dasein *is* so for essential reasons. Dasein's existence as temporal, which Heidegger outlines in great detail in *Being and Time* and contemporaneous works, provides the specific ontological constitution of being open to things, to others, to the world, to time, and to being itself. This temporality that grounds transcendence is what animals conceivably lack.

Heidegger writes that animals do not comport themselves toward things, but are rather taken by them, that they remain confined within their surrounding ring. They are unable to surpass and transcend other beings because they are fundamentally held within their own being. This captivation within the ring, it bears repeating, has nothing to do with a real barrier or limit that surrounds the animal. Dasein is not transcendent because it surpasses some boundary that animals are unable to surpass. Rather, transcendence is a primordial phenomenon of being itself, which animals purportedly lack. The following passage from Heidegger's 1928 course speaks to this very point in a way that prepares us for his comments on the animal's ring:

> Transcendence does not mean crossing a barrier that has fenced off the subject in an inner space. But what gets crossed over is the being itself that can become manifest to the subject on the very basis of the subject's transcendence. Because the passage across exists with Dasein, and because with it beings which are not Dasein get surpassed, such beings become manifest as such, i.e., in themselves.... Therefore, what Dasein surpasses in its transcendence is not a gap or barrier "between" itself and objects. (GA26, 211–12/166)

The phenomenon of surpassing accounts for Dasein's ability to let these other things be. Animals, in contrast, are simply too absorbed with their own inner drivenness to encounter things manifestly. In his essay "What Is Metaphysics?," Heidegger captures the necessity of transcendence for the encounter with beings as such: "Such being beyond beings we call *transcendence*. If in the ground of its essence Dasein were not transcending [e.g., if one were an animal], which now means, if it were not in advance holding itself out into the nothing, then it could never comport itself [*sich . . . verhalten*] toward beings nor even toward itself" (GA9, 12/91). This is precisely the position of animals, all wrapped up within their self-encirclement. They cannot comport themselves toward the world because transcendence is not a primordial fixture of their manner of being.

Working back toward the ecstatic character of time that grounds Dasein's transcendent constitution as being-in-the-world, the animal might now be conceived in the following way. Animals are unable to relate to other things *as such*. This, specifically, is a linguistic construction, but its roots delve beyond language. Animals lack language because language is essentially determined by transcendence. Transcendence is a character of being-in-the-world as always already being beyond oneself in being toward the world as something that happens. This ontological condition is meanwhile the foundation for Dasein's comportment toward things. Human Dasein can comport toward things, it can leap out of itself, because it is transcendent. Animals, on the other hand, have been described as well behaved insofar as they are fixed within their spheres. They do not escape, do not jump away, whereas, as Heidegger explains in a later course on Herder (and where Uexküll is referenced just before this note), the "human being is not bound to a determinate sphere: He is unbound, has freedom" (GA85, 138). This notion of ek-sisting beyond oneself is itself grounded in the ecstatic character of time. As we have already been told, numerous times throughout various works, Dasein is the temporal being as such. So the question that seems most pressing, and the one that requires some response, is whether animals are temporal beings as well. If they do not have an ecstatic character of time, then the rest of these pieces (transcendence, world, language) will fall into place for Heidegger's analysis. Either a coherent image of animal life will emerge or we will be presented with only the outline of pieces that together form a disrupted—dare I say shattered?—picture of animals.

For the most part, it is fairly well known how the being of Dasein is rooted in the ecstatic character of time. Time is the horizon on which the preceding analyses ultimately come to rest. Heidegger concludes as much in the final pages of *The Basic Problems of Phenomenology*: "transcendence, on its part, is rooted in temporality and thus in Temporality. Hence *time is the primary horizon of transcendental science, of ontology*, or, in short, it is the *transcendental horizon*" (GA24, 461/323). In order to better frame this discussion of animals and time, however, it will help to briefly revisit Dasein's temporal dimension, albeit in a manner specific to the territory of irruptions and borders that we looked at earlier.

Briefly, Heidegger is clear that Dasein is a transcendent being on the basis of time. "The *ecstatic character of time makes possible Dasein's specific overstepping character, transcendence*, and thus also the world" (428/302). It is this very capacity of Dasein to overstep itself that I consider of great importance because, by comparison, it is the animal's inability to step out of its encircled being that characterizes its being so vividly. We have already noticed that Heidegger emphatically rejects the connotation of some spatial boundary or limit that Dasein oversteps in transcendence. Similarly, the animal's ring is not a spatial or physical limit. Yet, there is a limit nevertheless, and it has to do with time. Indeed, Heidegger in fact claims that *there is* a line that Dasein must leap over in actualizing itself, but it is only a very thin line.

> Yet between this uttermost brink of possibility and the actuality of Dasein there lies a very fine line. This is a line one can never merely glide across, but one which man can only leap over [*überspringt*] in dislodging his Dasein [*seinem Dasein einen Ruck*

gibt]. Only individual action itself can dislodge us from this brink of possibility into actuality, and this is the *moment of vision* [*der* Augenblick]. (GA29/30, 257/173)

The limit that is surpassed is, as noted earlier, nothing 'out there'—or 'in there,' for that matter—but a limit of what one is capable of. Unlike the animal that cannot escape (*über . . . hinausspringt*) itself, Dasein can and must leap over itself. This moment of vision that jolts Dasein to itself is a possibility of time.

Earlier in his assessment of boredom, Heidegger offers a brief account of the *Augenblick*, just as he does in *Being and Time* (§65). A treatment of the moment of vision will have to wait for another reading, since all that I'm interested in now is how it relates to our look at animal life.[15] The as-structure has already been noted for its irruptive qualities between being and beings, and now what I wish to highlight is a more primordial irruption within Dasein itself. This rupture is associated with the moment of vision: "The moment of vision ruptures [*bricht*] the entrancement of time, and is able to rupture it, insofar as it is a specific possibility of time itself" (GA29/30, 226/151). The entrancement of time has to do with the temporal dimension of boredom (*Langeweile* translates literally as a 'long while'), and we will have reason to recall it in a moment. But what is the significance of this rupture? And what is revealed in Dasein's moment of vision? Most of all, it reveals Dasein's own temporal horizon:

> Why must that expanse of the entrancing horizon ultimately be ruptured by the moment of vision? And why can it be ruptured only by this moment of vision, so that Dasein attains its existence proper precisely in this rupture? Is the essence of the unity and structural linking of both terms ultimately a *rupture* [ein Bruch]? What is the meaning of this *rupture* [Gebrochenheit] *within Dasein itself*? We call this the finitude of Dasein. (252/170)

The seismic irruptions that we noted earlier are seen here again, just as active as before. What is more clear in this instance, however, is that it is Dasein's finitude, its temporal ek-sisting, that is more definitively claimed as the source of the ruptures. Moreover, the ecstatic nature of Dasein's time, ushered in by the moment of vision, ruptures what he names the "entrancement of time." In boredom, for example, time entrances Dasein (and, in time, Dasein is entranced with beings as a whole). But might this not suggest a similarity between Dasein's entrancement and an animal's captivation? Though there is the hint of a 'close proximity' between the two, Heidegger replies that their similarity is "merely deceptive" and that "an abyss lies between them which cannot be bridged by any mediation whatsoever" (409/282). The abyss is echoed a page later, but this time as an abyss within Dasein itself, undoubtedly arising from all the irruptions.

What we have, then, is a continual revisiting of ruptures and abysses: within Dasein, between Dasein and time, within time itself, between being and beings, and between Dasein and animals. Further, there is even that strange disruption within the essence of the animal itself. I purposely emphasize this terminology because it underscores the radical nature of Dasein's temporality. On the basis of the ecstatic dimension of time, Dasein transcends itself and others as being-in-the-world. There is a primordial rupture within the fabric of Dasein's being. And it is this temporal rupture that is the source for the leap that is denied to animal being. Françoise Dastur also remarks that "the guiding thread in the analysis of animality is therefore again a transcendental understanding of finitude."[16] This temporal dimension is the reason that the "*essential disruption*" within the essence of the animal proves so enticing.

It should really come as no surprise that the world is wrapped up in the cloak of time. It is from the temporal dimension that the world is revealed in Dasein's transcendence. This leads to the question of animals and time. Do animals have a sense of time? An initial response would be to say that surely they must. Even if it is only in terms of circadian rhythms that reveal times of day or the 'inner clocks' that tell Canadian geese, for example, to return northbound from their southern sojourn. But this is surely not the ecstatic character of time that interests Heidegger and fundamental ontology. The time of the animal is another question entirely. Despite his references to animals and organisms, Heidegger has surprisingly very little to say about them in conjunction with time. The closest he comes to formulating a direct connection is in *Being and Time*, with the following elliptical remark: "It remains a problem in itself to define ontologically the way in which the senses can be *stimulated* or *touched* [Reiz und Rührung] in something that merely has life, and how and where the Being of animals, for instance is constituted by some kind of 'time' " (GA2, 346/396). The scare quotes around "some kind of 'time' [*eine "Zeit"*]" provide a strong indication of what Heidegger's answer might be to this problem. The being of animals does not appear to be constituted by time, at least not by the ecstatic character of time that frames Dasein's existence as reported in *Being and Time*.

What are we to make of this curious passage? For the most part, Heidegger answers the first part of this problem in the 1929–1930 lecture course. The animal's senses are defined as touched or stimulated in Heidegger's ontological definition of the animal's being disinhibited by things within captivity.[17] But how and where is the being of animals constituted by some kind of 'time'? Derrida, for example, cites this passage, but, as elsewhere in his essay, he saves this problem for a later analysis that likely won't ever come.[18] He does offer one hint, however, by suggesting it is more than a coincidence that Heidegger poses this problem directly before a chapter entitled "The Temporality of Falling [*Die Zeitlichkeit des Verfallens*]." Heidegger's discourse is always centered around Dasein here, but the descriptions offer some degree of parallel with respect to his analysis of animal life, principally in terms of the body, the leap, and the moment of vision. It may be that the "present" body provides insight into the time of the animal.

To begin, note the similarities between the following comments on Dasein and the analysis of capacities that Heidegger delivers in 1929–1930:

> Like the concept of sight, 'seeing' will not be restricted to awareness through 'the eyes of the body.' Awareness in the broader sense lets what is ready-to-hand and what is present-to-hand be encountered 'bodily' in themselves with regard to the way they look. Letting them be thus encountered is grounded in a Present. This Present gives us in general the ecstatical horizon within which entities can have bodily *presence*. (GA2, 346/397)

We have not forgotten the main question behind this analysis: namely, how animals relate to their environment. This excursus on time is just another manner of approaching the relational character of life, and Heidegger reaffirms this here. He does so, moreover, with an uncharacteristic appeal to bodily (*leibhaftig*) presence. If an entity can present itself before Dasein, and do so bodily (which seems particularly important to me), it does so only from the basis of the ecstatic horizon of the present.

As ecstatic, we will be interested to discover that the present "leaps away [*entspringt*]" from an "awaiting" or "waiting toward" that is characteristic of time (347/397). The present leaps away from both its future and past in letting "Dasein come to its authentic existence only by taking a detour through that Present" (348/399). There are a number of distinctions that Heidegger makes throughout these sections on temporality, not the least of which is the one between the inauthentic present as "making present" and the authentic present in the "moment of vision." But since both 'presents' are of an ecstatic nature, this does not concern me so much as whether or not we can think of animals as even being present and being presented by something, let alone whether such presence is inauthentic or authentic. This is not a dismissal of this distinction; however, it is of less importance for us because, as *Being and Time* makes quite clear, even humans are for the most part inauthentic in their daily dealings, while nevertheless existing temporally.

The direction of this analysis into the ontology of animality leads one to wonder if, in their limited access to things, animals encounter things as present. Whether, that is, we can say the following concerning animals as Heidegger does of Dasein, namely, that "taking action in such a way as to let one encounter what *has presence* environmentally [*des umweltlich* Anwesenden], is possible only by *making* such an entity *present*" (326/374). My concern in this is that there must be some continuity underlying a living thing's relations to its surroundings. Even if animals have a limited access to what is open to them, the focus of this relation must be grounded in something. What binds a living being to its environment? For Dasein, it comes down to the horizon of ecstatic time. Do animals, therefore, have a similar foundation in time? In asking this, I share the same concern as the one Didier Franck poses in his essay "Being and the Living." Franck writes:

> But can the temporal constitution of life and the living be considered a separate, that is to say, in the end, a secondary problem? . . . Indeed, if the being of an animal were to be excluded from time, Being itself would thereby lose the exclusivity of its temporal meaning, and, if we live only by being incarnate in a body that testifies to our kinship with the animal, the ontological detemporalization of the animal would imply that the living incarnate that we are is existentially inconceivable, and that we must abandon the name of *Dasein*. (137)

Are animals excluded from time? Or is the being of the animal such that things are capable of becoming present?

Based on our earlier readings, our first impression should be that entities are not 'made present' before animals. In the encircling ring, things withdraw even in the openness of a disinhibited relation. Things do not appear as such. But does this imply a lack of temporality? William McNeill offers the clearest insight into Heidegger's language on precisely these points. In his reading of the 1929–1930 lecture course, McNeill indicates how nothing "enduring" can present itself to the animal in part because the animal itself attains no "permanence." After citing two compelling passages from Heidegger, McNeill writes: "In order for a living being to be able to achieve an endurance beyond or in excess of the temporal flow of presentation, it (its living presence) would have to take up an independent stance in relation to something outside of and beyond not only that which is presenting itself, but beyond the present of whatever is presenting itself at each moment."[19] He finishes this thought by noting that "such permanence is not atemporal or eternal, but an enduring in the manner of the specific temporality of *historical* time."

It may be too much to ask, therefore, that animals have a present, or that they live in the present. To be in the present suggests a break or gap between oneself and the other thing, to be able to comport oneself independently, which does not seem to be the behavior of animals. Rather than letting things be, animal behavior "obviates" the thing; rather than a fullness of presence, the animal environment resembles the "continual production of an emptiness" (GA29/30, 367/252). If this is the case, what are we to make of Françoise Dastur's comment on the temporal character of the organism? Dastur writes:

> It is not simply the case—no more than it is a concern with respect to humans, as Heidegger will later affirm in the "Letter on 'Humanism' "—of adding a soul to the animal organism, but rather to see in the organism something other than just a simple living body [Leib] as a presence simply given [le simple corps vivant (Leib) en tant que presence simplement donnée], and therefore to understand it as a "dynamic" phenomenon, that is to say, essentially temporal, of an organization becoming constant. (Heidegger, 53)

This passage is all the more provocative insofar as it appears in relation to Uexküll's biology. The organism is not a static being, even if it is not the "enduring" presence capable of standing over against its world. According to Heidegger's reading, a permanent (or Dastur's "constant") ontological state would require an "attending to" of things, a degree of endurance (Heidegger's term is Bleiben, which can also mean 'to remain') in this relation. For the animal, there appears to be no time like the present. Animals are incapable of comporting themselves over and against that which presents itself; they are always already taken by things, captivated, and driven, to the extent that they never achieve what McNeill calls "an independent stance" within their environment. The being of animals is always implicated in their being captivated by things, and thus they are never entirely at liberty to be their own self, even to be a self, beyond the "proper peculiarity" of being animal.

And yet Heidegger does suggest a temporal character within animal being. This arises earlier in his lectures when he is in the process of differentiating organs from instruments, specifically in how organs are morphologically bound up with the organism as a whole. Within this context, we discover another of Heidegger's infrequent appeals to animal temporality. Just as with many of his other comments on these difficult issues, Heidegger again flirts with our patience while opening a venue for more unanswered questions. It may also offer a glimpse of what might be meant by an "organization becoming constant" in the temporality of animal life.

> The organs as established features, as in the case of the higher animals, are bound to the lifespan [die Lebensdauer gebunden]

of the animal, i.e., not merely in the first place to time as an objectively definable period [*an die objektiv feststellbare Zeit*] during which the animal lives. Rather the organs are bound into and are bound up with the temporal span [*die Dauer*] which the animal is capable of sustaining as a living being. Even if we cannot pursue here the problem concerning this relationship of the organism and its organs to time [*zur Zeit*], it is already clear from such general reflections that *organ and equipment relate precisely to time in fundamentally different ways*. And it is this which first grounds an essential distinction in their respective manners of being, if we accept that the temporal aspect [*der Zeitcharakter*] is metaphysically central for each manner of being. (GA29/30, 328/224–25)

There are too many things to be said here. We could point to how Heidegger raises a seemingly central issue—how organisms and organs relate to time—only to not pursue it. We could notice how the character of time not only grounds the essential distinction between organs and equipment, but provides the first ground for each manner of being. Time appears as the horizon of an ontological difference within the behavior of living beings, but why is it not pursued further? Similarly, is there a temporal difference between animals, and not just between animals and Dasein? If so, how does this change the essential definition? This would not be a great leap since Heidegger already acknowledges a difference between the organs of "higher" animals and those of presumably lower animals. For example, the "structureless and formless" lives of "tiny protoplasmic creatures" have organs, but due to their lack of a firm shape, "their organs are therefore temporary organs" (327/225). But if Heidegger makes a distinction here between higher and lower animals (to say nothing of Dasein that stands outside of either category), then he is also making a temporal distinction, which might also mean an essential distinction in their manners of being. The organs are bound up within the temporal span of the animal, so, to the degree to which animals differ, we might have a different notion of time at play here. Not just as an "objectively definable time," but in terms of the life span of the organism as a whole. We could also point to the fact that the being of animals is constituted in these lectures, as opposed to *Being and Time*, by some kind of time, and this time without the scare quotes. Animals have a relationship to time, even one that might very well be essential to their manner of being.

Unfortunately for the animal, I fear that this problem of time comes back to the encircling ring again. Heidegger will later write that "this capability for encirclement is the fundamental characteristic of the animal's actual being in every moment of its life-span [*in jedem Moment seiner Lebensdauer*]" (375/258). If animals are temporal, they are so only to the extent that they are "within time," but not constituted by ecstatic time as such. The entire span of their lives is characterized first and foremost by the behavioral distinction of captivation; their captivation precludes a proper relation. Unlike Dasein, therefore, for whom a world *is* insofar as Dasein temporalizes itself (GA2, 365/417; GA24, 383/271), the animal does not temporalize itself. Their lives are framed by an inability to leap out of themselves, to transcend beings, to adopt a stance ("*Haltung*," which is linked to "comportment [*Verhaltung*]") over and against these things, an inability therefore to be in a world and to be able to say as much. Animals might be touched by time, but they themselves do not have that shock or irruption that might allow them to touch ecstatic time in return.

What does this leave us with? This question is interestingly left unresolved. Despite all indications that Heidegger gives to warrant our abandonment of a more fundamental understanding of time in the case of animals, he intentionally returns to the open-endedness of these issues. Every sign suggests that he has made a final decision, that he is going one way and not the other, but then he circles back to a point where the matter is left incomplete. For example, in a section entitled "The incompleteness of our present interpretation of the essence of the organism," Heidegger notes that some biologists have started to refer to organisms as historical beings (such as Theodor Boveri in *The Organism as a Historical Being*). If animals are born, mature, age, and 'die,' then this might also suggest that they are historical. "What kind of history [*Geschichte*] do we find in the life process of the particular individual animal? What kind of history does the animal kind, the species, possess? . . . Can we and should we speak of history at all where the being of the animal is concerned? If not, then how are we to determine this motility?" (GA29/30, 386/265–66). This section offers a novel direction for pursuing our thoughts on the being of animals, one that will spill over—or leap—into our look at the ontologies of Merleau-Ponty and Deleuze. I'm thinking specifically of how motion enters into the fray. Heidegger's look at the essence of animals has offered a number of interesting insights into the lives of organisms, not the least of which is his trademark analysis of time. In pointing out that his analysis is incomplete, he at the same time points toward "a task that we are only now just beginning to comprehend." One gets the sense that something new, something not yet explored, is beginning to take shape. And it has to do with motion: "All life is *not simply organism* but is *just as essentially* process, thus formally speaking *motion*" (385/265). It is as though he is at once speaking of the history of the animal as ontic being, and also suggesting something altogether different from the concept of "organism." Life as process, life as motion, rather than the life of a being.

He further notes: "Captivation is not a static condition, not a structure in the sense of a rigid framework inserted within the animal, but rather an intrinsically determinate motility which continually unfolds or atrophies as the case may be" (385–86/265). I don't mean to suggest that Heidegger is already implying the "process ontology" that Deleuze will develop, or that Heidegger does away with the organism as an ontical unit. The concepts of process and motion, however, do point toward a new vista for rethinking the ontology of the animal. Perhaps this phenomenon of life that one calls "organism" *is* more dynamic than previously thought. A new plane emerges, and it has to do with the body, movement, and processes.

AN AFFECTED BODY

The comparative analysis that Heidegger undertakes in 1929–1930 is more or less an exclusive one. While he will return to the being of animals over the course of his life, he will never again do so with such attention. It is all the more compelling, therefore, that this comparative analysis results in more questions to be asked and more problems remaining unanswered. One can even question the relevance of thinking of this analysis as a "comparative" one insofar as there is next to no comparison between human Dasein and animal life, ontologically at least. Ontically, Heidegger never denies that there is room for biological or physiological comparison, that humans and animals are entities everything else being equal. But this doesn't get to the essence of the matter.

Just what is the nature of the animal's relation to the environment? Something that I have not raised until now is that whenever Heidegger refers to an animal's surroundings, it is almost always written as "*Umgebung*" rather than as "*Umwelt*." Animals, it seems, are not only deficient in Welt; they are also poor in Umwelt. The concept of Umgebung implies the surroundings, environment, habitat, or milieu, within which the animal is encircled, but it does not have the immediate connection to the world that Umwelt does. It is unsurprising, therefore, that Heidegger will write, in a later lecture course, "World is always *spiritual* world. The animal has no world (Welt), nor any environment (Umwelt)" (GA40, 34/47). Does Heidegger change his position between 1930 and 1935? Similarly, in the 1946 "Letter on 'Humanism,'" Heidegger reaffirms his hesitation by noting that animals have their Umgebungen, though not the Umwelten that Uexküll popularized. Leaving aside the question of language, the following remark demonstrates just how "puzzled" Heidegger remains on the question of living things: "Because plants and animals are lodged in their respective environment [Umgebung] but are never placed freely into the clearing of being which alone is 'world,' they lack language. But in being denied language they are not thereby suspended worldlessly in their environment [*Nicht aber hängen sie darum . . . weltlos in ihrem Umgebung*]. Still, in this word 'environment' [*Umgebung*] converges all that is puzzling about living creatures" (GA9, 157–58/248). Such is the state of his thought some seventeen years on from the time of his comparative analysis. The relation to an environment is still the puzzling domain of animal life. There is still a glimmer that animals are not worldless like the rock, but they are likewise not placed freely in the domain of an *Umwelt*, let alone the clearing of world.

It is the flip side of this analysis of animality that has received the most attention in previous readings. Critiques of Heidegger's account of Dasein came early, especially from those close to him. Jonas and Löwith, both former pupils, questioned the degree to which human Dasein was elevated above the rest of nature. More recently, Derrida has claimed that there can be "no animal *Dasein*" for Heidegger (*Of Spirit*, 56), and Franck, a bit more emphatically, states that "never in the history of metaphysics has the being of man been so profoundly disincarnated" (146). But while attention has been paid to recontextualizing the bodily, and even animal, dimension of human Dasein, I am more interested in Heidegger's accounts of animal behavior as ontologically significant. From early on, Heidegger shows a familiarity with biological writings—and a preference for Uexküll in particular—but though his analyses suggest a proximity to the key issue of the environment, his language says otherwise. It is not the *Umwelt* that he attributes to the animal, but the more innocuous *Umgebung*. Animal behavior, as compared to human comportment, reveals the ontologically decisive measure by which the natural world is carved into human existence and animal life, with that ever so slight abyss gaping between them. Human Dasein has made the leap across this chasm, while the animal remains immured in its own captivity. But it is not entirely alone, since it is on the same ontological divide as rocks, trees, angels, and God, a notion that Deleuze would later appreciate in his own ontological equation of the tick with God. The animal may show signs of disruptive behavior, but a kind of shocked behavior is all it is. Nevertheless, Heidegger's conclusions suggest the intrinsic necessity of how the animal's body behaves as a structural relation, and possibly a temporal one at that, even if the relation is to nondescript surroundings.

> There is no indication that the animal somehow does or ever could comport itself toward beings as such. Yet it is certainly true that the animal does announce itself as something that relates to other things, and does so in such a way that it is *somehow affected* by these things. I emphasize this point precisely because this *relation to . . .* which is involved in animal behaviour, even

though it essentially lacks the manifestness of beings, has either been quite overlooked in previous attempts to define the concept of the organism and the essence of the animal in general, or has merely been inserted as an afterthought. (GA29/30, 368/253)

By 1967, in a course conducted with Eugen Fink on Heraclitus, Heidegger will hold that the body was always the most problematic issue. "*Fink:* Back in the day when you first came to Freiburg, you said in a lecture course: the animal is poor in world. At that time you were on the way toward the kinship of human beings with nature. *Heidegger:* The phenomenon of the body is the most difficult problem" (GA15, 234/146).[20] The course that Fink mentions is of course that of 1929–1930, a course that Heidegger would subsequently dedicate to Fink on its preparation for publication. The body is a problem; so too is the organism and environment. Their discussion continues: "*Fink:* The only question is how 'organism' is to be understood here, whether biologically or in the manner that human dwelling in the midst of what is is essentially determined by bodiliness. *Heidegger:* One can understand the organism in Uexküll's sense or as the functioning of a living system. In my lecture, which you mentioned, I have said that the stone is worldless, the animal poor in world, and the human world-forming.... The human body is not something animalistic. The manner of understanding that accompanies it is something that metaphysics up till now has not yet touched on" (234/146). Nearly forty years after the fact, it is Uexküll that Heidegger still associates with animal life and against whose idea of the animal he still contrasts human being. Uexküll has clearly provided a formidable companion in Heidegger's engagement with biological life. He suggests that the body still hasn't really been properly considered, a remark that we will remember Uexküll making as well. There is a convergence of themes, then, between the puzzling domain of the environment, the difficult problem of the body, and the incomplete analysis of how life's movement and process enmesh with being. Merleau-Ponty will conceptualize this bodily relation and he will do so in conversation with Uexküll. It is to his ontology that we now turn.

CHAPTER 4

The Theme of the Animal Melody
Merleau-Ponty and the Umwelt

Without Heidegger's knowing it, the untouched problem of the body that he mentions in 1967 had already received a voice in "metaphysics." In a working note to *The Visible and the Invisible*, published posthumously in 1963, Merleau-Ponty expresses "why I am for metaphysics": "For me the infinity of Being that one can speak of is *operative*, militant finitude: the openness of the Umwelt—I am against finitude in the empirical sense, a factual existence that *has limits*" (VI, 305/251). Merleau-Ponty's aspiration for metaphysics is in part to reconceive our understanding of the world not in comparison to the infinite or eternal (the Unendlichkeit) but in terms of what he repeatedly refers to as the openness (Offenheit) of the Umwelt (e.g., VI, 222/169; 238–40/185–86; 250/196; 266/213). We are not to understand this as a contradiction of Heidegger's explicit determination of animal Umwelten as closed; Merleau-Ponty is not responding directly to Heidegger's theses on animal and human being. In fact, there is no indication that he ever knew of Heidegger's theses or of his engagement with Uexküll. Rather, in his late thought Merleau-Ponty entertains a return to ontology rooted in nature whereby being—what he will call "wild" and "brute" being—reveals itself in the interstices of the body and the world. The openness of the Umwelt, and not the infinity of the world, is the hidden source and ontological horizon of the embodied animal subject.

His metaphysics, if we still wish to call it that, conceives of the body not as a thing, substance, or essence, but as an unfolding relation to an Umwelt through the phenomenon of behavior (N, 270/209). The question is therefore that of understanding how the body relates to its environment, what this reveals of the ontology of nature, and how this addresses that prickly issue of the human and animal. All of these issues involve

Merleau-Ponty's recurrent analysis of behavior as a means of expressing the living body, from his earliest writings on *The Structure of Behavior* through to his final manuscripts left incomplete at the time of his death in 1961. His focus provides us with a rich view in our mounting consideration of onto-ethologies.

There are a few issues that I wish to highlight in this chapter. The first is Merleau-Ponty's appeal to Uexküll's biology. As was the case with Heidegger, Merleau-Ponty's writings on organisms and the structures of life remain invested in the work of Uexküll, particularly in his consideration of the *Umwelt*. The appearance of Uexküll is further associated with the more general development of Merleau-Ponty's later ontology, one that is noticeably rooted in the being of nature. Such an ontology rests on the significance of movement, and how the organism's bodily behavior melds with the "flesh" of the world.

THE STRUCTURE OF BEHAVIOR

After remaining overlooked for many years, *The Structure of Behavior* is again finding a captive audience.[1] Among those contributing to this renewed interest are scholars who have sought to retrace the development of Merleau-Ponty's thought, particularly owing to the many questions that surround the fragmentary nature of his late writings. *The Visible and the Invisible*, Merleau-Ponty's famously incomplete work due to his untimely death, provokes many of these questions, not the least of which is due to the over one hundred pages of "working notes" that elliptically trail off into the intellectual project of surmising Merleau-Ponty's 'unfinished' thought.[2] Not unlike Mozart's *Requiem* or Bach's *The Art of the Fugue*, *The Visible and the Invisible* leaves us wanting more, hanging, as we do, on the final notes of a compelling yet incomplete score. This text has left scholars pondering where this work was headed, what may have lay in store for Merleau-Ponty's thought, and how he might have sought to reconstitute what he was calling, in his final years, "the new ontology." The restoration of Merleau-Ponty's final work has thankfully not overshadowed its tremendous implications: Merleau-Ponty *was* in the process of working out a new direction in his phenomenology, and it was a direction that sought to borrow from many of the themes that one finds in his earlier thought, such as those involving the concepts of the body, behavior, perception, nature, being, and the world. However, the reader also discovers many new and enticing concepts as well, such as the flesh, chiasm, brute being, the importance of art and language, and the increasing insistence on a new ontology. But this new direction, I believe, was not a break from his past. One does not discover a "turn" in his thought, no more than one finds a renunciation of his earlier tenets.³ Even if one finds certain self-criticisms, such as the popular one regarding his early dependence on consciousness—the often quoted note "The problems posed in *Ph.P.* are insoluble because I start there from the 'consciousness'–'object' distinction" (VI, 237/200)—these point to no more of an explicit turn than do, for example, Nietzsche's self-criticisms of his earlier works. Rather, I am inclined to agree with Renaud Barbaras who sees a gradual "evolution" in Merleau-Ponty's thought, and with Martin Dillon's assessment that "there is not so much a turn as a development in Merleau-Ponty's thought during the last fifteen years of his life. That is, I see a continuity in his thinking rather than a leap to a new position; I see modifications rather than reversals."⁴

Added to the incomplete version of *The Visible and the Invisible*, we could also note the appearance of new publications of Merleau-Ponty's lecture notes from the courses he offered at the Collège de France throughout the 1950s. These lectures have been appearing increasingly in France throughout the 1990s, and have started to appear recently in English translations. These publications contribute a further source of academic novelty, for they provide a clearer depiction of Merleau-Ponty's development, particularly in how he works, from the years after the *Phenomenology of Perception*, toward an elucidation of an ontology of nature. An example of such a publication, and one that will be central to our exposition, is the three lecture courses that Merleau-Ponty delivered in the late 1950s that have appeared fortuitously even if fragmentarily: 1956–1957's "The Concept of Nature," 1957–1958's "The Concept of Nature: Animality, the Human Body, and the Passage to Culture," and 1959–1960's "The Concept of Nature: Nature and Logos: The Human Body." These lectures provide a valuable source for reconstructing the development of his later thought as well as for reconsidering the importance of his earliest work on animal behavior. All of this is to say that there has been good reason to return to Merleau-Ponty's more neglected early work. While the *Phenomenology of Perception* has always remained central to the scholarship of his phenomenology, this has not always been the case with his earliest publication. With our attention firmly oriented toward understanding the contours of Merleau-Ponty's ontology of the animal, we will begin with an examination of this early thought before passing on to the developments of his final works.

The theme of the organism and nature is evident from *The Structure of Behavior*'s opening sentence. "Our goal," Merleau-Ponty writes, "is to understand the relations of consciousness and nature: organic, psychological or even social" (SB, 1/3). One of the primary reasons our attention is drawn to this relation is because Merleau-Ponty finds an inadequate—or, at the very least, doubtful—relation between the naturalism found in science and the transcendental critiques of philosophy. On the one hand, the tendency of science to lean toward naturalism has resulted in an overly empiricist account of life, one that relies on physico-mechanical causes to "imply a nature in itself." On the other hand, one confronts similar problems from less physical, more idealist domains, such as biology's vitalism, psychology's dependency on the mind, and the idealist strain of transcendental philosophy. In contrast to this mix, Merleau-Ponty evokes a third alternative to describe the natural world: namely, by beginning with a "neutral" analysis of behavior. By beginning in this fashion, Merleau-Ponty highlights how he is "starting 'from below,' " as if to suggest that a study of behavior will reveal a more profound ontological basis for the emergence of nature. Though he does not develop the metaphor of depth here, it can be read in conjunction with the indifferent ontological pretensions of physics that he notes in his introduction, as well as his use of the term "archeology" to describe the later ontological writings on nature. In *The Structure of Behavior*, behavior acts as a supposedly neutral catalyst for digging beneath the gloss of empirical nature while at the same time avoiding the positing of an idealist force. We will see that in the late 1950s he will again appeal for an act of digging, but this time as an archeology of nature to get at the brute being beneath perceptions.[5]

Of greatest interest for our present study is how the organism is characterized within this early discussion of behavior. With respect to contemporary biology, Merleau-Ponty notes that the study of behavior presents a novel position to the two established trends of mechanism and vitalism, and he does so in a manner different from Heidegger. Both mechanism and vitalism are said to be theories that "remain open"—that is, they are not closed off and thus not surpassed or overtaken—largely due to their adherence to a "realistic" view of life. The problem with both options, however, is that "our picture of the organism is still for the most part that of a material mass *partes extra partes*" (SB, 1/3). In contrast to the theory of nature that Merleau-Ponty will develop, and particularly one that will appeal explicitly to "structure," this picture of the organism evokes a strong disparity. To think of the organism as a "material mass *partes extra partes*" is to dissociate the living being from any relational structure. What Merleau-Ponty refers to as "our picture" is certainly not *his* picture, but rather the common scientific view that reflects, whether intentionally or not, a view of the organism as belonging to an atomistic universe. Cut off from the environment in which it lives, every organism can be defined, classified, dissected, and studied as a "material mass" existing external to everything else around it. This, we will recall, is precisely the interpretation of the world and entities that Heidegger so stridently opposes before entertaining his own postulation of the being of the animal. There are no intrinsic relations; just material mass abutting against material mass, each isolated from the other insofar as each is its own self-contained unity.

But, in his emphasis on behavior, Merleau-Ponty appears susceptible to the very same charges brought against "behaviorism," the field of psychology pioneered by Ivan Pavlov, John Watson, and B. F. Skinner that focuses on the externally observable patterns of animal life. In other words, how does Merleau-Ponty's emphasis on behavior abandon the atomistic and strictly physiological approach favored by those who famously espoused the field known as behaviorism? Here too, Merleau-Ponty draws exception to another familiar distinction, namely, that between the "mental" and the "physiological" in psychology. The problem with "behaviorism" is that it merely continues to uphold an "atomistic interpretation" of the organism, albeit disguised under its new name. In behaviorism, he notes, "behavior is reduced to the sum of reflexes and conditioned reflexes between which no intrinsic connection is admitted" (3/4). For Merleau-Ponty, one of the implicit tasks is to restore meaning to the concept of behavior. It is "neutral," he states, precisely because it does not take sides between the mental or physiological theories of organisms. Indeed, rather than rejecting both the mental and the physiological outright, Merleau-Ponty's version of behavior takes both sides as he seeks to unite a physical view of life with the reintroduction of consciousness. This unity is emphasized as one of "structure," a term that borrows heavily from the Gestalt theory of "form" as it emphasizes the whole of the organism as being more than just the sum of its parts. There is something unique in the structural whole of a living being that cannot be reduced to its various organs, fluids, appendages, cells, reflexes, and so on. Even if we were to interpret the organism as the accumulation of diverse parts, this would amount to no more than resubmitting a mechanistic view of life. The organism would just be a whole added up through the accumulation of its parts, just as a machine is the totality of all of its gears, levers, and parts. Therefore, there must be something about the organism that is irreducible to an atomistic interpretation but that does not also slip in a vitalist life force. Merleau-Ponty finds an initial answer in his look at the structure of behavior, particularly as he weaves between the mechanism and vitalism of biology, between the physical and the mental of psychology, and between the empiricism and intellectualism of philosophy.

Behavior, therefore, is far from any old characteristic. Rather, it is already evident that it has a special affiliation to the essence of organisms. In a footnote, we read that "one says of a man or of an animal that he behaves; one does not say it of an acid, an electron, a pebble or a cloud except by metaphor" (2/225 fn.3). Perhaps an obvious point but a pertinent one no less, and all the more so when we later consider Deleuze's ontology. For Merleau-Ponty, behavior is descriptive of the organism as a whole, with the understanding that behavior cannot be attributed to either an organism's organs or to other independent things that may nevertheless exhibit signs of movement, such as a cloud or an electron. Although the cloud moves, as does an electron or pebble, they remain caught within a causal mechanism such that movement is dependent solely on external determinants. Likewise, Merleau-Ponty will critique the theories of "reflex behavior" as positing a similar causal relation between stimuli and reactions, whether it is between the organs themselves or between the organism and environmental stimuli. Neither form of causality is suitable for capturing the meaning of behavior since both ignore the totality of the organism's being. At issue, then, is how to capture the meaning of the organism's totality without reinvesting in a mechanical or vitalist program, inasmuch as each relies on separated parts acting on one another in either mechanical relations or as an entelechy. We have already observed a similar stance in Heidegger's writings, so where and how they differ on the point of behavior will be important.

The phenomenological interest in mereology, and specifically with respect to the organism, will remain with Merleau-Ponty throughout his writings. For example, in the *Nature* lectures he emphasizes the importance of the parts–whole distinction: "How are we to understand this relation of totality of parts as a result? What status must we give totality? Such is the philosophical question . . . at the center of this course on the idea of nature and maybe the whole of philosophy" (N, 194/145). If we are to find the totality of the organism in the phenomenon of behavior, then we must inquire into its meaning.[6] But to look for an explicit definition of behavior may be an ineffective path. Merleau-Ponty doesn't so much offer a clear formulation inasmuch as he offers a variety of different views. Accordingly, behavior is linked with a variety of other notions: behavior as structure, behavior as form, behavior as signification, behavior as a manner or attitude of existing. In each case, however, it is the ontological interpretation of the organism that is conveyed. Behavior demonstrates a relational enclosure insofar as the organism is structurally united with its world. This allows Merleau-Ponty to escape two antithetical views: on the one hand, the atomism of being a substance *partes extra partes*, and, on the other hand, the deceptive trap of simply introducing the psychophysiological notions of "integration" or "coordination" to link up the organism as a whole (SB, 84/76). The problem with interpreting an organism's behavior as the integration or coordination of its diverse parts is that it too hastily determines the organism's form as one pertaining solely to its inner parts. Neither atomism nor integration constitutes a totality of the organism, whose forms "are defined as total processes whose properties are not the sum of those which the isolated parts would possess" (49/47). As he will later note, "The genesis of the whole by composition of the parts is fictitious" (53/50; cf. 163/150).

It is here that behavior assumes its relevance. Behavior is not "a thing" nor is it an empty "idea"; rather, "behavior is a form" (138/127). It is a form, moreover, that executes a "higher" relation between an organism and its surroundings, uniting the two in an unprecedented way. "The relations of the organic individual and its milieu are truly dialectical relations therefore," Merleau-Ponty writes, "and this dialectic brings about the appearance of new relations" (161/148). Above all, this appearance of new relations takes place within the context of a world as the ontological vista for all organic behavior. For this to be so, there must be a prior and fundamental relation to the world out of which all other relations may be considered. The world—or environment or milieu, since Merleau-Ponty has not yet distinguished between these concepts—emerges as a Gestaltist framework from out of which a picture of the organism may present itself.

When read in this way, there is no question that Merleau-Ponty is influenced by Heidegger's treatment of the concept of world. Heidegger is not mentioned by name until the final pages of the book—and he is referenced as an open question on the very issue of the world—but his ontological elucidation of world certainly underlies Merleau-Ponty's thought throughout. Consider the following claim, where behavior is directly linked to the animal's being and world:

> The gestures of behavior, the intentions which it traces in the space around the animal, are not directed to the true world or pure being, but to being-for-the-animal [*l'être-pour-l'animal*], that is, to a certain milieu characteristic of the species; they do not allow the showing through of a consciousness, that is, a being whose whole essence is to know, but rather a certain manner of treating the world, of "being-in-the-world" [*être au monde*] or of "existing." (137/125)

Behavior is not invoked here as the key that will open our eyes to "the true world or pure being," whether for the animal or for us. There are no unrealistic expectations in this regard. Instead, behavior offers us something far more exciting—namely, our means of accessing the mode of being-animal, which, importantly, is expressed as a manner of being-in-the-world. This relation between an animal's behavior and the world is a reciprocal one, each being dependent on the other. As Merleau-Ponty claims, it is "a truly dialectical relation," for just as much as behavior reveals the being of the animal as found in the world, the world is equally uncovered in the behavior of the animal. "The world," Merleau-Ponty writes, "inasmuch as it harbors living beings, ceases to be a material plenum consisting of juxtaposed parts; it opens up [*il se creuse*] at the place where behavior appears" (137/125).[7]

Just as Heidegger critiqued the metaphysical view of the world as composed of present-at-hand entities, Merleau-Ponty also aims to circumvent a theory of the world that posits self-sufficient and closed material entities. The world is not composed of isolated substances (e.g., atomism) that may or may not link up with one another in a purely external manner (e.g., integration and coordination). Nor is the world the entirety or sum of all extant entities. Rather, it is through the phenomenon of behavior that an alternative view presents "as an expression of totality,"[8] such that the animal shows itself in its *being*. In the case of a chimpanzee, for example, we are told that it may only be "a short and heavy manner of existing" (138/126)—the chimp can stand upright but not always, it can grasp boxes but only as a tactile object—but it is an existing in the world nevertheless. As we have just observed, behavior both traces the animal-being as "being-in-the-world" and it carves out an opening on to the world itself. It elicits, therefore, an ontological dimension of the animal in which the animal's totality converges with that of its milieu. This perspective leads Merleau-Ponty to claim that "the organism has a distinct reality which is not substantial but structural" (139/129).

Thus, what we see in Merleau-Ponty's early investigations is a philosophy of structure and form being substituted for a philosophy of substance. Nowhere is the emphasis on form more clear than in his omnipresent use of Kurt Goldstein's 1934 study on the "so-called holistic, organismic approach" to the living organism.[9] If there is one constant source on which Merleau-Ponty draws in *The Structure of Behavior*, it is Goldstein's unique method of uniting biological, psychological, and physiological studies into a unified Gestalt theory of the organism. Goldstein's new method broached an important way of treating the human being as a whole, particularly with respect to the neurological and/or physiological disablement of individuals. Goldstein's work was largely conceived as a response to his treatment of soldiers suffering from a variety of ailments brought on by war conditions. As a physician, he found that most approaches to "abnormal" functioning relied too heavily on a fractured view of the human; none of the theories permitted an adequate view of the person as a whole. With this in mind, it is clear how Merleau-Ponty might be influenced by such a novel approach to the living being, especially in association with the Gestalt psychology of Kurt Koffka and Wolfgang Köhler. But as much as Goldstein figures into Merleau-Ponty's thought, it is the appearance of another theorist that I'm particularly taken by. Jakob von Uexküll does not appear often in the pages of *The Structure of Behavior*, but his thought is at least partially evident and will continue to remain with Merleau-Ponty when he again takes up the question of organic life in his lectures on nature.

Uexküll's appearance possibly owes something to Goldstein himself, for Merleau-Ponty will surely have read the short section that Goldstein pays to Uexküll's theory of the "tonus valley" (*The Organism*, 89–90). But, even more directly, it may have been the following passage that led Merleau-Ponty to look more closely into Uexküll. Goldstein writes:

> We must make a clear distinction between the surrounding world in which the organism is located, and the milieu that represents only a part of the world—that part that is adequate to it, that is, that allows for the described relationship between the organism and its environment. Each organism has its milieu, as Jakob von Uexküll has emphasized. Its existence and its 'normal' performances are dependent on the condition that a state of adaptation can come about between its structure and the environmental events, allowing the formation of an 'adequate' milieu. (105–106)

Such a relation between the organism's structure and its milieu *may* have led Merleau-Ponty to Uexküll, but there is no firm proof for this. Indeed, Uexküll is never cited as such in *The Structure of Behavior*, and he won't be addressed until the second course on *Nature*. It may be just as likely that Merleau-Ponty owes his introduction to Uexküll to the writings of Buytendijk, the Dutch biologist who also featured in Heidegger's 1929–1930 lectures, whom Merleau-Ponty does cite.[10] If a connection to Uexküll isn't yet obvious, it becomes more definitive in the sole reference made by Merleau-Ponty to Uexküll in *The Structure of Behavior*, whom he quotes only through Buytendijk: " 'Every organism,' said Uexküll, 'is a melody which sings itself' " (SB, 172/159).

It is not so much the fact that Uexküll appears in Merleau-Ponty's thought—and at the moment, he does so only secondhandedly—that is of interest. The historical, biographical, and archival dimensions of this lineage is certainly interesting, particularly as it also concerns the appearance of Uexküll's biology in the thought of Heidegger and Deleuze.[11] But no matter how interesting these relations may be, this is not my primary interest. Instead, I am more interested in what Merleau-Ponty *does* with Uexküll's biology and, in this respect, how his usage of Uexküll differs from that of Heidegger and Deleuze.

We discover in Merleau-Ponty's flirtation with Uexküll a particular manner of expressing the being of the organism. If we are to understand the organism as a totality and as a structure that exceeds its physiological body, then the question is how best to express the intimate relation between the organism and environment as a single form. How can the organism be ontologically expressed such that one emphasizes the structural and relational dynamic without reasserting a substantialist or mechanical view of the organism? I believe that Merleau-Ponty finds his language at least partially in Uexküll's elucidation of animal *Umwelten*. The first sign of this is in the only claim attributed to Uexküll that I noted earlier: "Every organism is a melody that sings itself." The musical motif is barely a theme in this early work, but it is important enough to Merleau-Ponty, specifically in its application to the structural relation between organism and environment. It is also to this musical theme that he will return in his *Nature* lectures.

What can we therefore discover in Merleau-Ponty's use of musical metaphors? To begin, on separate occasions Merleau-Ponty makes independent musical references to both the world and the organism. For example: "the world, in those of its sectors which realize a structure, is comparable to a symphony, and knowledge of the world is thus accessible by two paths: one can note the correspondence of the notes played at a same moment by the different instruments and the succession of those played by each one of them" (SB, 142/132). Despite the appeal in thinking of the world as a symphony, however, we are instead confronted with the possibility that the world, if known in this way, might relapse into an interpretation whereby the symphony is really only the summation of all of the various notes. This would imply a conglomerate world according to a mechanical view: each note from each instrument creates the symphony as a whole, but a whole that is just the sum of its parts. This clearly won't work with respect to the direction of Merleau-Ponty's train of thought, as he himself is aware.

From another point of view, we find a comparable remark concerning the organism, and this time it is a consideration of the organism as the instrument of music. The analogy seems to work, at least at first glance: if the world is a symphony, then living beings could be the instruments producing this music. Might it then be possible to think of the organism as a natural "keyboard"? One could imagine a multitude of keyboards, each producing a countless stream of melodies. But this analogy ultimately doesn't work either: "The organism cannot properly be compared to a keyboard on which the external stimuli would play and in which their proper form would be delineated for the simple reason that the organism contributes to the constitution of that form" (11/13). As opposed to a keyboard, which can be played only by external stimuli, organisms actively contribute to the melody itself. In other words, an organism is not a passive instrument that is excited and stimulated in a reactive manner but a form that sings itself.

This would mean that the organism is actively engaged in the 'playing' of itself as well as its environment. The environment and organism are intimately related in some musical theme, though not one that can be reduced to merely external stimuli. In the end, the world is not just a symphony and organisms are not just keyboards, where each is dependent on the other in only an external determination. In order to formulate the nature of this structural relation, Merleau-Ponty appeals to the *Umwelt* for the first time, and he does so through a revealing citation. He cites Goldstein as claiming that "the environment emerges from the world through the being or actualization of the organism. Stated in a less prejudiced manner, an organism can exist only if it succeeds in finding in the world an adequate environment—in shaping an environment" (*The Organism*, 85; cf. SB, 12/13). Despite the attraction of this claim, Merleau-Ponty nevertheless hesitates on this relation, believing that it still posits the organism as merely 'offering' its keys to the environment, thus not entering a truly equal relationship. Although I think he misinterprets Goldstein slightly—or, at the very least, doesn't yet take seriously the environment as the creation of the organism—it is nevertheless an indicative reference for a couple of reasons. Firstly, it binds the organism and environment in an active sense. What I mean by this is that he treats the behavior of the organism as reciprocated by the environment: behavior is described as an "effect" of the organism's milieu (the environment forces the organism to behave in particular ways), but the milieu is also already established by the preceding behavior of the organism (the environment appears as it does due to an initial act of behavior). Each is locked together in the movement of the organism through its environment, though not as symphony and keyboard. The second aspect of this reference is its affiliation to Uexküll. The quotation on the *Umwelt* that Merleau-Ponty pulls from Goldstein appears just a short paragraph after Goldstein names Uexküll's research as "so generally valid that it no longer meets with much opposition" (84). This association further captures the importance of Uexküll's thought to Merleau-Ponty, even if comes via Goldstein. Uexküll's silent appearance is particularly striking insofar as he seems to underlie many of the important claims made with respect to this relational dynamic. What awaits further study, therefore, is how the organism's *Umwelt* may be reconsidered according to a different musical theme.

As opposed to the metaphors of a symphony and keyboard, Merleau-Ponty finds something more appealing in the expression of a "melody." Unfortunately, the concept of the melody is not formulated as such within *The Structure of Behavior*, but the term does weave its way through his thought in such a manner that is hard to ignore. More than anything else, the notion of melody is used to express the unity of the organism as a whole and as a theme that finds its rhythm flowing through the environment as well. The notion of a melody, in other words, appears to be Merleau-Ponty's manner of explicating the relational structure of the living being as such. A sense of this can be seen in one of his attempts to reinterpret the meaning of 'coordination.' Considering that he was previously critical of a particular understanding of coordination (i.e., coordination as the antithesis of atomism), it is not especially helpful that he maintains the same term to convey his own interpretation of an organism's unity. Nevertheless, he is clear that he is offering a "very different" type of coordination, and it is one that is expressed in terms of melody:

> Here the coordinated elements are not only coupled with each other, they constitute together, by their very union, a whole which has its proper law . . . just as the first notes of a melody assign a certain mode of resolution to the whole. While the notes taken separately have an equivocal signification, being capable of entering into an infinity of possible ensembles, in the melody each one is demanded by the context and contributes its part in expressing something which is not contained in any one of them and which binds them together internally. (SB, 96/86)

At first it seems that Merleau-Ponty, despite his intentions, has reintroduced a mechanical construction of the organism insofar as the whole is still a product of interconnected notes. However, unlike a machine, this form of unity is dependent on neither a prior construction in which each part (or note) is necessarily connected to another, nor is this unity a material one. The melody, by contrast, signifies the organism as a whole where each of its parts resound through the entirety of the organism, though not in any predefined way. Each note captures the whole. In an otherwise unremarkable sentence, Merleau-Ponty later makes the same claim, but this time in comparison to a soap bubble. As we will recall, the soap bubble is a favorite analogy of Uexküll's. "In a soap bubble [*une bulle de savon*] as in an organism, what happens at each point is determined by what happens at all the others" (141–42/131). This is not an innocent analogy. If we were not already aware of Uexküll's inclination to compare the *Umwelt* to a soap bubble, Merleau-Ponty's analogy may have been quickly passed over. Yet the soap bubble in this passage functions in the same manner that the melody does, and it cannot be by coincidence that both are found in Uexküll's thought. More important, however, both melody and soap bubble express the unity of the organism as a reverberating totality.

This unity, moreover, is not material, despite the physical existence of the organism. The parts fit together with one another, but the unity "is not a simple consequence of the existence of organs or substrate. The process of excitation forms an indecomposable unity and is not made up of the sum of the local processes" (97/88). It is perhaps for this reason that the idea of a melody ultimately proves to be more illustrative than a soap bubble, which still retains a physical connotation. But what sustains this melody? When Merleau-Ponty writes that "coordination is now the creation of a unity of meaning which is expressed in the juxtaposed parts, the creation of certain relations which owe nothing to the materiality of the terms which they unite" (96/87), what holds this balance together? What is the "unity of meaning" of which he speaks that creates the structural relation? Does Merleau-Ponty, in eschewing materialism, open the door to a vitalist life force?

Despite these possibilities, the irreducible quality of the organism's being does not reinvest in a vitalist force either. The organism's unity is not confined solely to its bodily apparatus due to its behavioral activities in an environment. Thus, the organism is not solely a physical specimen because it *is* itself only in its inherent relations with its milieu:

> We are upholding no species of vitalism whatsoever here. We do not mean that the analysis of the living body encounters a limit in irreducible forces. We mean only that the reactions of an organism are understandable and predictable only if we conceive of them, not as muscular contractions which unfold in the body, but as acts which are addressed to a certain milieu. (164/151)

The 'immaterial' aspect of the organism, if we may call it that, shows itself in the manner that living beings engage with their surroundings. Their environments contribute to their totality, but do not yet 'complete' them. Accordingly, it is with this relational dynamic that the importance of the melody comes into play.

Another way of putting this is that the structure of the organism—taken as a whole, as a form—pushes one to another level of relation. Each structure is only the node for many intersecting relations, almost approximating, albeit quite loosely, the intersecting lines of Deleuze and Guattari's strata and assemblages: "The form itself, the internal and dynamic unity which gives to the whole the character of an indecomposable individual, is presupposed by the law only as a condition of existence . . . the existence of such a structure in the world is only the intersection of a multitude of relations—which, it is true, refer to other structural conditions" (153/142). That the organism is a whole (*as* form, *as* structure) means that it by necessity relates to still other structures. And by this, Merleau-Ponty largely means the environment.

Despite the frequent references to the environment, Merleau-Ponty seldom makes a strong distinction in his variable terminology for the world, at least not in comparison to the nuanced usage of language that we observe with Heidegger, who is at pains to distinguish between Dasein's world and the environments of animals. In many respects, Merleau-Ponty glosses over this distinction in all but one explicit place. The environmental world of organisms—which is alternately expressed as "*ambiance*," "*entourage*," and more frequently, "*milieu*"—contributes to the organism's being, but, as was the case with the organism and its parts, the organism is not merely a part within the world as a whole. "Science," writes Merleau-Ponty,

> is not therefore dealing with organisms as the completed modes of a unique world [*monde*] (*Welt*), as the abstract parts of a whole in which the parts would be perfectly contained. It has to do with a series of "environments" [*ambiances*] and "milieu" (*Umwelt, Merkwelt, Gegenwelt*) in which stimuli intervene according to what they signify and what they are worth for the typical activity of the species considered. (139–40/129–30)

We have already seen a similar claim made by Heidegger concerning the objectivity of the world and we have also observed Uexküll's critique of physics for its one-world view. Likewise, Merleau-Ponty does so here by invoking the significance of individual environments through a reference to Buytendijk, who is himself utilizing Uexküll's terminology of *Umwelt, Merkwelt*, and *Gegenwelt*.[12] Every living being does not live within one 'unique' world, but within a specific *Umwelt* that is significant to it. This relation, moreover, is again what I believe Merleau-Ponty conceives as a melodic construction.

However, instead of continuing to press into the domain of defining the ontological conditions for the organism's relation to an environment—and, in so doing, expressing further the notion of the melody—Merleau-Ponty's discourse slides into an account of how *we* might perceive this relation. In other words, his thought moves from the relation between organism and environment to what conditions *our perception* of this relation. One could say that he more or less stops addressing the organism as a living being in favor of taking up the "the perception of the living body," what he will call the "phenomenal body" (169/156). This does not mark a new direction within his text, for a phenomenological approach has been evident from the start. It is just that the phenomenology of perception becomes more pronounced—beginning with the third chapter "The Physical Order; The Vital Order; The Human Order"—when Merleau-Ponty moves away from his critique of previous paradigms and begins laying the groundwork for his own contributions.

To this end, we are no longer dealing with what might be an ontological interpretation of the organism and its inherent relation to an environment, but with how this relation is constructed out of our own perceptive lives. It is we, Merleau-Ponty notes, who form the relation through perception. Thus "form," "structure," "melody," and "meaning," all important characteristics for the being of the organism, give way to new formulations, where form, structure, melody, and meaning are modes of perceptual knowledge of the world in which organisms appear. For example, "It should not be concluded from this that forms *already* exist in a physical universe and serve as an ontological foundation for perceptual structures" (156/144). To be sure, form is *not* a physical thing existing within the world. But rather than offering a clarification of this "ontological foundation," we instead learn that "form is not a physical reality, but an object of perception" (155/143). Discussion slides away from an ontological foundation toward how *we* might perceive this foundation. In conjunction with the notion that form does not just pertain to the organism but to the perception of the organism as a form, Merleau-Ponty makes similar remarks concerning the structure of life as one imminent to consciousness. Consider, for instance, that one must "describe the structures of action and knowledge in which consciousness is engaged," particularly since "the problem is still to understand how the objects of nature are constituted for us" (178/164–65). In part, they are constituted to us according to a melody and rhythm. But whereas the melody may have been at one time something uniting the organism and its environment, we are led also to understand the melody as something that has a rhythm for our knowledge: "there are melodic unities, significant wholes experienced in an indivisible manner as poles of action and nuclei of knowledge" (179/165–66). This is not the melodic unity or significant whole that belongs to the organism in its relational being. The melody and significance of the whole derives from our *perception* of this structure. Indeed, toward the end of his analysis, "structure" (as well as "meaning") is defined only in terms of its association with our perception: "*structure*" is "the joining of an idea and an existence which are indiscernible, the contingent arrangement by which materials begin to have meaning in our presence" (223/206). Thus, the attempt to express the living being becomes an investigation into how we perceive the living being.

In many respects, this direction in Merleau-Ponty's thought is evoked in the same passage in which he raises Uexküll. " 'Every organism,' said Uexküll, 'is a melody which sings itself.' " After citing this formulation, however, Merleau-Ponty does not continue to dig into this metaphor. Instead, he continues: "this is not to say that it knows this melody and attempts to realize it; it is only to say that it is a whole which is significant for a consciousness which knows it, not a thing which rests in-itself" (172/159). The melody, which may have once been an interesting means for developing an ontology of the organism, gives way to its significance for conscious perception. This is not to say that Merleau-Ponty abandons the view that organisms sing the melody themselves. Nor that they are united with specific *Umwelten*. Nor even that an organism is a "unity of signification." It is just that he, in the words of Renaud Barbaras, "suspends" his inquiry into "natural being" in favor of the domain of consciousness in which the natural world comes to be perceived.[13] For the present, it suffices to notice that the organism as a melody that sings itself never receives fruitful exposition. It indicates a novel dimension for pursuing the ontology of a living being, though one that never seriously entertains the nature of this melody. The suspension of natural being shall be removed in his later lectures to which we will turn shortly.

To close this initial discussion, I find that Merleau-Ponty's early position with respect to living beings can be fairly summarized with the following passage. While still seeking to extract his position from the mechanist and vitalist distinctions, Merleau-Ponty writes: "to understand these biological entities . . . is to unite the ensemble of known facts by means of their significations, to discover in all of them a characteristic rhythm, a general attitude toward certain categories of objects, perhaps even toward all things" (SB, 171/158). Signification, rhythm, and attitude are each laden with meaning, and each refers to the being of the organism. However, the organism itself recedes into the background in order to focus on how human consciousness unites this ensemble of facts. But even if Merleau-Ponty does turn away from the organism itself, he has laid the groundwork for a future study of nature. Fortunately, the theme of the animal melody will again emerge as important to the ontological unity of organism and environment.

A PURE WAKE, A QUIET FORCE

One such indication of continuity between Merleau-Ponty's *The Structure of Behavior* and his late work on nature and ontology is the emphasis on the relation. With the concept of "structure," the notion of relationality almost goes without saying. It is at once a structure of the organism as well as the structure uniting conscious perception with the thing perceived. One can find another, and not unrelated, passage that holds much in anticipation of certain themes that will be announced in the late 1950s. I am thinking primarily of the importance of the "flesh" and Merleau-Ponty's development of the "cleavage" between the body and the world. Toward these concepts, we find a foreshadowing in his early text: "Life is not therefore the sum of these reactions. In order to make a living organism reappear, starting from these reactions, one must trace lines of cleavage [*des lignes de clivage*] in them, choose points of view from which certain ensembles receive a common signification and appear" (SB, 165/152). I would like to suggest that Merleau-Ponty conjures precisely such a reappearance of the organism in his

Nature lectures, and that he does so through tracing, if you will, the lines of cleavage between the living organism and the environment it inhabits.

In so doing, Merleau-Ponty returns to a certain terrain that he departed from in *The Structure of Behavior* and more immediately in *Phenomenology of Perception*. Within these works, as partially discussed earlier, Merleau-Ponty is concerned with describing the manner by which conscious perception apprehends the world. But rather than advancing consciousness as just another transcendental version of a disembodied cogito, he appeals to the lived body as the source of a "rootedness" within the world. The embodied world, and particularly in terms of the self-movement of the body, unites the organism's relation with its being in the world. For example, he remarks that "our bodily experience of movement . . . provides us with a way of access to the world and the object . . . which has to be recognized as original and perhaps as primary. My body has its world, or understands its world, without having to make use of my 'symbolic' or 'objectifying function'" (PhP, 164/140–41). Though this emphasis on the living body ushers in a new method of conducting phenomenology, it is also the case that Merleau-Ponty's phenomenology remains implicated in a philosophy of consciousness. On this point, Renaud Barbaras notes in *The Being of the Phenomenon*: "its [PhP] sole mistake is that it remains on the descriptive level, being content with bringing to light this domain which still must be thought out. This revision, which will ultimately consist in passing from a description of the perceived world to the philosophy of perception, will be the objective of later works" (17). What Barbaras and others latch on to is Merleau-Ponty's own recognition that he had not yet given an ontological account of the perceived world. His earlier thought drew too much on the nature of a subject's conscious perception of the world in which it lives, as opposed to providing the foundation for the nature of perception itself. It is the nature of the world as original and primary that now becomes the main focus.

As noted, Merleau-Ponty was aware of the new direction set before him. It is on this point that we discover a further deepening or burrowing in the trajectory of his thought. In a working note to *The Visible and the Invisible* he writes: "Results of *Ph.P.*—Necessity of bringing them to ontological explicitation" (VI, 237/183). And again, just to emphasize the direction of his thought: "Necessity of a return to ontology—The ontological questioning and its ramifications: the subject-object question, the question of inter-subjectivity, the question of Nature" (219/165).

Nowhere is this task of unfolding a "new ontology" more explicit than in his final works. His aim from the outset of *The Visible and the Invisible* is to invoke a prescientific understanding of the world in the language of ontology, in terms of what he defines as *"the meaning of being"* (33/16). To better carry out his investigation, he introduces his now famous formulations of "flesh," a term that has "no name in traditional philosophy to designate it," as it is neither material nor spiritual, neither matter nor mind, and thus can be no better elucidated in ontological terms than as "a sort of incarnate principle that brings a style of being wherever there is a fragment of being. The flesh is in this sense an 'element' of Being" (184/139). In order to derive an account of this element of being he must first reexamine how the body can better lead us to understand, in a prereflective fashion, "our living bond with nature" (34/17). But it is precisely the status of this "living bond" that is always in question throughout this work: is Being a world that truly underlies everything like "universal flesh"? If so, how can a body, when "intertwining" and "blending" with other lives, in its "openness upon the world" (57/35), in its "prepossession of a totality which is there before one knows how or why" (65/42), be thought and located along an individual plane? That is to ask, how can the cohesive relation, this "prelogical bond" between living bodies and things, not be implicated in an organic or vitalist model that presupposes an all-encompassing unity that Merleau-Ponty ultimately does not wish to uphold? And further, how does this bond with nature either add to or depart from Heidegger's descriptions of animals as bound and captivated, whereas being human is unbound?

More specifically, Merleau-Ponty must be able to respond to the notion, on the one hand, of "a cohesion without concept, which is of the same type as the cohesion of the parts of my body, or the cohesion of my body with the world" (199/152), while, on the other hand, state that "surely there does not exist some huge animal whose organs our bodies would be, as for each of our bodies, our hands, our eyes are the organs" (187/142). While there is a lot at stake in Merleau-Ponty's treatment of the living body in conjunction with other things of the world, it is the nature of this cohesive bond, or this "cleavage," that really addresses the tangible density of relationships between bodies, both living and nonliving. What Merleau-Ponty clearly wants to avoid is a descriptive account that would posit a world as having a life in and for-itself, independent of the living beings that constitute it. Thus, in claiming that there is a preconceptual cohesiveness between body and world, he does not wish to claim that this cohesion implies a natural life (an "animal" life, to be more specific) that our individual bodies belong to in the manner that our organs belong to our bodies.[14] There is a split then, a chiasm if you will, in the fabric of the natural order. This gap apparently distinguishes between certain living bodies, such as between my body and that of an other, but on another level there *is* a cohesion between living things, one, however, that does not amount to a great living force, the world as one huge animal and we its organs.

And yet, as Merleau-Ponty asks, "[w]here are we to put the limit between the body and the world, since the world is flesh?" (VI, 182/138). How is it that there can be a cohesion between the organs of a body and between this body and the world, but not to the extent that this cohesion, the flesh of the world, would obtain something like an organic life unto itself? In other words, how can the world be universal flesh, but not understand the flesh as a 'life' independent of the bodies that constitute its totality? Furthermore, how does the cohesive relation express a unity within nature that leads Merleau-Ponty to speak of a "new ontology"? Does this ontology bear on the difference, if any, between animal and human?

The nature of this cohesion between organic bodies, and between bodies and a world, is already under way in his lecture courses on *Nature*. In the second course, "The Concept of Nature, 1957–1958: Animality, the Human Body, and the Passage to Culture," Merleau-Ponty never really made it beyond the first topic on animality. It is here that he most explicitly addresses the work of Uexküll—who he teased us with in *The Structure of Behavior*—to better elucidate how organisms produce and come to have a cohesive relation to their environments in a decisive manner. Within this course, Uexküll's *Umwelt* proves to be an evocative manner for expressing this structural relation. Indeed, as one commentator has remarked, and I'm led to agree, "this chiasm is the philosophical payoff of Merleau-Ponty's interpretation of Uexküll's work."[15] I therefore have two primary goals in the remaining section, and both are interrelated. The first is to observe what Merleau-Ponty gains from his reading of Uexküll and, correlatively, to see how this reading may contribute to an elucidation of the flesh of the world as the basis for his return to ontology.

After having begun by reviewing certain limitations of Cartesian metaphysics—most notably, as a scientific paradigm enlisted to describe the essence of nature's beings—Merleau-Ponty addresses nature as a theme in need of ontological clarification. To this end, he picks up on precisely the same footing with which he began some fifteen years earlier, with the concept of behavior. But this time his analysis of animal behavior is oriented more explicitly by an engagement with modern biology than it was previously, where behaviorism was the greater target (N, 187/139–40). It is in this manner that Uexküll's theories initially appear. Well, it is perhaps misrepresentative to suggest "theories," for it is really only the concept of the *Umwelt* that takes pride of place in Merleau-Ponty's analysis. The description of the *Umwelt* is first set off against the notion of an objective or scientific world (*Welt*) that exists in itself. By comparison, the *Umwelt* is not only said to be the "purely subjective domain" of animal life, but, more pertinently, it is "the environment of behavior" (220/167). This may be the first sign that even though Merleau-Ponty appears to be reviewing Uexküll's thought before subjecting it to a philosophical interpretation, he is nevertheless already imparting his own specific reading to these descriptions. It is unsurprising, therefore, that the *Umwelt* is compared to behavior as well as to consciousness: "Uexküll anticipates the notion of behavior," Merleau-Ponty writes. "This behavioral activity oriented toward an *Umwelt* begins well before the invention of consciousness . . . . Consciousness is only one of the varied forms of behavior" (220/167). More than anything else, consciousness is only a type of behavior, which supports the view that all organisms, even at the level of embryos, exhibit some behavioral patterns even though they do not necessarily demonstrate any signs of consciousness. Otherwise put, the *Umwelt* underlies the possibility of consciousness and, as such, an organism's *Umwelt* provides a more profound and universal depiction of the living being. Insofar as Merleau-Ponty aims to circumvent the priority of conscious perception, a theory of the *Umwelt* may thus prove beneficial.

To this end, it is suggested that the unity of an organism "must rest on an activity" (224/170) that simultaneously unites the organism as a whole and acts as a cohesive bond between the organism and its *Umwelt*. This suture does not occur after the fact, but is ontologically constitutive of behavior itself. Movement is therefore central to our understanding of the organism, as Merleau-Ponty explains: "Between the situation and the movement of the animal, there is a relation of meaning which is what the expression *Umwelt* conveys. The *Umwelt* is the world implied by the movement of the animal, and that regulates the animal's movements by its own structure" (230/175). In this guise, movement is therefore a means of reconsidering how we understand the animal and the world as a cohesive structure. We are offered an evocative illustration of this, as Glen Mazis has shown,[16] when Merleau-Ponty describes the movement of a bird in flight in *Phenomenology of Perception*:

> If we want to take the phenomenon of movement seriously, we shall need to conceive a world which is not made up only of things, but which has in it also pure transitions. . . . For example, the bird which flies across my garden is, during the time that it is moving, merely a grayish power of flight and, generally speaking, we shall see that things are defined primarily in terms of their 'behavior' and not in terms of their static 'properties.' It is not I who recognize, in each of its points and instants passed through, the same bird defined by explicit characteristics, it is the bird in flight which constitutes the unity of its movement. (PhP, 318/275)

This "flurry of plumage" is a beautiful depiction of a "pure transition," a movement that evinces the unitary phenomenon of an animal with its environment through behavior. With this example, we begin to see how movement provides an opening onto Merleau-Ponty's onto-ethology, such that behavior is the locus for this transitory state that is also the site of a new phenomenon. The way in which the bird in flight manifests itself as a "unity" evokes a living being giving expression to itself. It holds itself together as a fold of nature. Later in the *Nature* lectures, Merleau-Ponty will explore further ways to arrive at an ontological expression of life, and he will do so in a manner that reveals his Bergsonian background as well as anticipates Deleuze's ontology of the actual and intensive. Consider the way he moves from a description of "unity" to one of "adhesion":

> It is less of the multiple in the living than of an adhesion between the elements of the multiple. In a sense, there is only the multiple, and this totality that surges from it is not a totality in potential, but the establishment of a certain dimension. From the moment when the animal swims, there will be life, a theater, on the condition that nothing interrupts this adhesion of the multiple. It is a dimension that will give meaning to its surroundings. (N, 207/156)

The behavior of the bird in flight, this flurry of plumage, becomes the adhesion of the multiple, the sustaining of a dimension expressive of life. But each dimension of life is only a momentary adhesion held together through behavior until interrupted by some other adhesion of the multiple. The bird-in-flight encounters scurrying-brown-mouse. The oily-otter-swimming emerges from the water to become slow-basking-otter. In each case, the animal-environment is transformed and takes on new meaning.

The focus on behavioral activity leads us toward a general depiction of life itself. More specifically, however, the activity and movement of organisms shed light on the natural cohesiveness between living beings that itself leads us toward Merleau-Ponty's ontology. Beginning with the general view,

> we must understand life as the opening of a field of action. The animal is produced by the production of a milieu, that is, by the appearing in the physical world of a field radically different from the physical world with its specific temporality and spatiality. Hence the analysis of the general life of the animal, of relations that it maintains with its body, of the relations of its body to its spatial milieu (its territory), of inter-animality either within

> the species or between two different species, even those that are usually enemies, as the rat lives among vipers. Here two *Umwelten*, two rings of finality [*anneaux de finalité*] cross each other. (227/173)

The interlacing of fields is suggestive, for it evokes the same penetration of *Umwelten* that was observed in Heidegger's analysis of the encircling rings. The different *Umwelten* of different organisms cross one another like rings 'opening' each to the other and giving the appearance of life itself. If life is opened up at all, it opens through such fields of action, which are more or less synonymous with an *Umwelt*. Even more striking is the notion that organisms are produced *by* the production of its milieu. The selection of Merleau-Ponty's phrasing is unambiguous: there is a reciprocal—and one may even say passive—relation between the organism and its milieu. The animal is said to be produced by the production of the milieu, but, in saying this, both animal and milieu are produced by a production that goes unnamed. Neither one is individually the producer, while both together are a product. What then produces the animal-milieu structure?

This is precisely the question that Merleau-Ponty is led to ask. Three or four times he attempts to formulate a question that will adequately articulate this relation: "How then does Uexküll understand this production of an *Umwelt*?" and again "But what is the subject that projects an *Umwelt*?" (231/176). An answer to these questions is of the greatest importance because it directs us toward the relational dynamic, and the adhesion of the multiple, within and between animals and their environments. Eventually we arrive at an interesting response, and it appears as a concept that we have already observed at work in his earliest text, namely, that of the melody. We should not really be surprised to find Merleau-Ponty returning to the same metaphor that he discovered in Uexküll fifteen years earlier, though thankfully he now finds reason to describe it at slightly greater length. If there is a production implicating both organism and environment at once, it might best be described as the "unfurling [*déploiement*] of an *Umwelt* as a melody that is singing itself." Merleau-Ponty continues: "This is a comparison full of meaning. When we invent a melody, the melody sings in us much more than we sing it; it goes down the throat of the singer, as Proust says. Just as the painter is struck by a painting which is not there, the body is suspended in what it sings: the melody is incarnated and finds in the body a type of servant" (228/173). The expression of this melody further reiterates the passive connotation of this existing production. By formulating the relation in this manner, Merleau-Ponty is clear to sidestep the possible misapplication of a causal determination existing between the organism and environment. Neither is the cause or effect for the other, but rather, as he notes of the melody, it "sings in us much more than we sing it." The melody seems to swell up through living beings without any voluntary or determinist implications, nor does the melody suggest the role of a higher reality, such as that associated with pantheism or *Naturphilosophie* discussed earlier in his course. In a manner of speaking, the *Umwelt* is a consistency of the relational dimension itself, which unfurls through living bodies like a melody. Subsequently, we may understand why Merleau-Ponty says "we must dissociate the idea of an *Umwelt* from the idea of substance or force" (231/176). The *Umwelt* is rather the "surging-forth of a privileged milieu," "a milieu of events," from within which the animal appears "like a quiet force," though unlike any vitalist life force (232/177). A further attempt to express this relation: "the unfurling of the animal is like a pure wake that is related to no boat" (231/176).

What I believe is at work within this philosophical interpretation of Uexküll's biology is Merleau-Ponty's grappling with "something new: the notion of *Umwelt*," though in a way in which he has not yet formulated a language for himself to express this relation. From one of his working notes, it is clear that a reconsideration and overhaul of his language are at work. "Replace the notions of concept, idea, mind, representation with the notions of *dimensions*, articulation, level, hinges, pivots, configuration—" (VI, 277/224). It is also clear that an ontological relation is at play, and that the relation involves neither substance nor force. The *Umwelt* unfurls like a melody, the animal unfurls like a pure wake. Adhesion of the multiple starts to sound pretty good but is never really developed. How can one describe this relation then? Instead of substance or force, the natural relation relies on a "melody," "a pure wake," or even a "surging-forth," each expressing the union of organism and environment simultaneously, but without appealing to mechanist or vitalist assumptions. The different attempts to explain this new dimension indicate that Merleau-Ponty has not yet settled on an adequate terminology, though it is very clear that something new is at work. The melodic element of being could be characterized by a "tangible density," a phrase used by Alphonso Lingis to describe the thickness of the world.[17] For how else might we explain the manner by which the organism's body remains "suspended in what it sings?" The melody is also described as incarnated (*s'incarne*), giving dimensionality and a certain thickness or density to the *Umwelt*. The *Umwelt* has texture, adhesiveness; it acts as a hinge, a dimension, a melodic production—indeed, I'm tempted to say that it is simply an "element" of being, though this carries a special meaning and would too quickly equate the melodic *Umwelt* with Merleau-Ponty's writings on "flesh."

In a few years' time, in his final course on *Nature*, we discover that Merleau-Ponty will move more easily from an account of Uexküll's *Umwelt* and the living body to a *"theory of the flesh"* (270–71/209).[18] But he does not do so here, in 1957–1958. A formulation of the flesh of the world, the sensible, the intertwining, and chiasm will have to await further introduction. For all this, however, the melodic *Umwelt* is not without its ontological dimension, as Mauro Carbone notes in *The Thinking of the Sensible*. Carbone draws attention to a type of "negativity" that underlies Merleau-Ponty's interpretation of the *Umwelt*, the melody as a theme that "haunts" the organism (37; cf. N, 233/178). Twice Merleau-Ponty mentions the haunting of the *Umwelt* melody, as 'something' that is present but only as an absence, as a life structure that for the moment resists appellation, as the composition of an environment that unfurls in the behavioral movements of the animal but that are never seen as such. Carbone aligns the theme of the melody with *"the absent,"* and this association is particularly appropriate when read in view of Merleau-Ponty's ontology where, for example, "totality is likewise everywhere and nowhere" (N, 240/183), both present and absent. Another way of expressing this dynamic in life is to note a relation between the visible and the invisible, a theme that would soon be omnipresent in Merleau-Ponty's thought. In the meantime, the notion of the melody will have to suffice to capture the meaning of the *Umwelt*. Merleau-Ponty concludes as much in the final paragraphs of his reading of Uexküll: "In brief, it is the theme of the melody, much more than the idea of a nature-subject or of a suprasensible thing, that best expresses the intuition of the animal according to Uexküll" (233/178).

A LEAF OF BEING

In the editorial notes to *The Visible and the Invisible*, Claude Lefort documents that Merleau-Ponty's manuscript dates as early as March 1959 (the "working notes" date from January) and continues through to the time of his death in 1961. This text was therefore in the process of being composed concurrent with when Merleau-Ponty was delivering his final lecture course on nature (1959–1960). The parallel between these two projects—really it is just one project, since "Nature" fits into the overall development of his incomplete return to ontology—is evident from the beginning of the final course, where themes that were pursued in the two earlier courses now merge with a more explicit orientation toward an ontological formulation. This formulation receives its impetus in nature, where Merleau-Ponty pursues "Nature as a leaf or layer of total Being—the ontology of Nature as the way toward ontology" (N, 265/204).[19] In pursuing nature as ontological, Merleau-Ponty is clear that he is not aiming to offer an epistemology or metaphysics of nature. It is natural being that he is after, and in particular the manner by which nature shows itself in the intertwining of lives, what he refers to as "inter-animality," or in the folds and leaves of being itself. What he is after, in other words, is what might be discovered in the "hollow" of being that remained unexcavated in his earlier works. He wishes to retrieve the "brute" or "wild" being that lies beneath all the cultural sediment of the intelligible world. As he notes, "there is no intelligible world, *there is the sensible world*" (VI, 267/214). So as opposed to continuing to describe the perceived world and the modes of perception, Merleau-Ponty intends to dig beneath perceptual consciousness in order to discover what allows for the possibility of perception itself. To do so, all the "bric-a-brac" associated with the cultural and intelligible world—"*Erlebnisse*," "judgments," "consciousness," "ontic" things—need to be observed for what they are: realities that have been "carved" out of "the ontological tissue" (VI, 307/253; cf. 324/270). It is not so much an issue of removing these aspects of life as much as digging back into the brute being of nature. Renaud Barbaras captures this intention: "He no longer takes consciousness as his starting point, which led him immediately to the problem of the relationship between the perceived world and nature; he begins with nature to show the identity in it of being and being-perceived. Thus, it is indeed by the reflection on nature that the transition towards ontology comes about."[20]

It is not that Merleau-Ponty no longer upholds a phenomenology of perception, but that perception and the perceived come to have a new sense in their application to natural being. Most of his focus highlights a series of divergences (*écarts*) that exists between things, but that, *as* difference, unites nature within the texture of being. What is particularly striking about his thought during these last few years is the language that he invokes to capture the leaves of natural being. Consider, on the one hand, some of the terminology to express *divergence*: chiasm, cleavage, folds, leaves (*feuillets*), invisibility, hinges, pivots, fields, and layers. There are even "fields of fields" (VI, 225/171), the folds may be "doubled, even tripled" (N, 275/212), or there may be a "whole series of layers of wild being" (VI, 232/178). Such folds in being may prove especially poignant when later compared with Deleuze's ontology. For the moment, the accumulation of layers manifests the texture of the divergence that exists within nature. Now consider, on the other hand, the terminology used to speak of the *unity* of such divergences: one reads of intertwining, the flesh, of its thickness, texture, fabric, pulp, the sensible, cohesion and adhesion, sutures, and seams. All of this language breathes a sensuousness into natural being, such that one can't help but feel an affinity for the cohesiveness of all things. Indeed, as Merleau-Ponty says at one point, it is not about "a hard nucleus of being, but the softness of the flesh" (N, 302/238). Yet what are we to make of this language? More specifically, what is suggested by Merleau-Ponty's late ontology with respect to Uexküll and the *Umwelten* of organisms?

The question that one really ought to ask here is: what is the ontological meaning of the expression "the world is flesh"? As the touchstone of Merleau-Ponty's late philosophy, the flesh is the element in which lives are lived. By calling the world flesh, Merleau-Ponty is certainly not speaking about human skin, though skin is also used as a parallel to the element of which he speaks. Rather, as "an 'element' of Being," the flesh is as if an extra dimension has emerged in addition to the traditional elements of water, air, earth, and fire. But the flesh is not just any old material or spiritual thing. Rather it is the element that makes being possible. It is the cohesiveness itself of the world, such that the world is rendered possible. One must keep in mind that Merleau-Ponty is attempting to describe natural being prior to conscious perception and intelligibility. This is "brute" or "wild" being; it is being in the hollows of the world of reflection, values, and thought. The prereflective, preconceptual, prespiritual *Vorhabe* of being, all find expression in this one carnal being of the flesh. The flesh, then, is tantamount to the sensible insofar as each alludes to the medium of what *is*. As already noted, "*there is* the sensible world": "The sensible is precisely that medium in which there can be *being* without it having to be posited. . . . The sensible world itself in which we gravitate, and which forms our bond with the other, which makes the other be for us, is not, precisely qua sensible, 'given' except by allusion" (VI, 267/214). Hence Merleau-Ponty's reluctance to subscribe to any one positing of this medium. There *is* flesh of the world, the sensible world.[21] They are but allusions to the *il y a* of wild being.

But if one cannot speak of the world but by allusion—"*one cannot make a direct ontology*" (233/179)—how do we find our way there? How does this ontological foundation become visible? Merleau-Ponty's thought does not stray that far from the insights he develops in *Phenomenology of Perception*; it is just the perspective that changes. As before, he continues to take the body as oriented in topological space as his cue for understanding the medium of being. But while he was interested earlier in how the body perceives the surrounding world, it is now more an issue of describing how the world *is* such that the body may be located within it. It is in this manner that Merleau-Ponty hastens to describe topological space as an especially relevant "model of being," particularly when compared to the harsh lines of Euclidean and Cartesian space. As opposed to the geometry of *res extensa*, Merleau-Ponty's view of the world is, by contrast, full and thick, "a total voluminosity which surrounds me, in which I am, which is behind me as well as before me" (VI, 264/210; N, 275/213). A world that is configured by the contours of a three-dimensional body exudes this sense of being replete. "I am open to the world because I am *within* my body" (N, 279/217), writes

Merleau-Ponty. The topological manner of being bodily is what allows for an initial penetration into the sensible being of the world's flesh.

Nowhere is the fullness of the sensible more evident than in Merleau-Ponty's descriptions of the touch of one hand touching the other hand.[22] This example—where one hand, say the right, touches the left hand while it touches back—in effect captures the reversibility of the subject and object in one and the same body, and even in one sole organ such as the finger (VI, 314/261). The one hand touches the other simultaneous with the other hand's touching, such that each reverses the relationship with respect to one another. Both are subject to the other's object to the extent that subject–object becomes a meaningless distinction, at least as defined by their traditional parameters. This prereflective sensation proves to be an example of intercorporeity in that it applies not only between two hands touching, but also extends to the perception of other bodies and the corresponding relations between two bodies. Merleau-Ponty writes:

> this is also an opening of my body to other bodies: just as I touch my hand touching, I perceive others as perceiving. The articulation of their body on the world is lived by me in the articulation of my body on the world where I see them. This is reciprocal: my body is made up of [aussi bien fait de] their corporeality. (N, 281/218)

As he describes elsewhere, I see myself seeing because I see others perceiving me. There is the visible for me only because I am myself possessed by the visible; I am *of* the visible (VI, 177–78/134–35). Correspondingly, it is never the case that there are "things" or *blossen Sachen* to be experienced in themselves. We do not live in such an intelligible world of positivist or present-at-hand entities. Rather, "our most natural life as humans intends an ontological milieu" (S, 206/163) where the world is invested with meaning due to the reversibility of experience, as expressed by the hand touching touch, and the divergence such reversibility implies, namely, the separation or chiasm between the hands. The *écart* is just as important as the relation itself, for, as Merleau-Ponty notes, "the touching is never exactly the touched" (VI, 307/254), and I never fully succeed in either touching myself touching or in seeing myself seeing. As a living being, one is constituted through the flesh of the world but there is just as much an "escape [from] *oneself*" (303/249) due to the invisibility in the visible and the untouchable in touch. In this respect, the touch might be distinguished from the intellectual grasp associated with total possession of oneself and/or the object.

The world is flesh because our bodies are themselves of the flesh, as the thickness that exists between the two hands, or between the perceiver and the thing, or one body and an other. In the end, the reversibility of things announces the need to rethink the ontological foundations of such relations: "it is imperative that we recognize that this description also overturns our idea of the thing and the world, and that it results in an ontological rehabilitation of the sensible" (S, 210/166–67). Such is the nature of the leaves, layers, and folds in Merleau-Ponty's thought. The flesh or the sensible is ultimately our navigation into understanding the structure of being, as it is both the union and divergence between all things: "It is because there are these 2 doublings-up that are possible: the insertion of the world between the two leaves of my body[,] the insertion of my body between the 2 leaves of each thing and of the world[.] This is not anthropologism: by studying the 2 leaves we ought to find the structure of being" (VI, 317/264). The flesh of the world emerges within these two leaves of the body, because the body is the chiasm of being sensible and being sentient; the body is the "sensible sentient" (179/136–37). The two leaves of the body are either side of the hinge in the becoming-one of the body's subject–object; the body is a two-in-one, both sensing and sensed, both separation and cohesion, visible and invisible. Such is also the case between more than one body, where the separation (chiasm, divergence, etc.) between bodies implies just as much their unity (intertwining, cohesion, etc.): there is "a surface of separation between me and the other which is also the place of our union, the unique *Erfüllung* of his life and my life" (287/234). The world is in the body just as much as the body is in the world.

But even to note that the body has two leaves seems too contrived for Merleau-Ponty, too flat and too oppositional. Instead, this notion might be better expressed as segments of a circular whole that captures the body in its behavioral movement. It is here that Uexküll once again becomes important to our interpretation of Merleau-Ponty's new ontology. For instance, we discover in the third course on *Nature* that Merleau-Ponty repeats the need for a rehabilitation of the sensible, and this time in direct reference to Uexküll. "This being-there by difference and not by identity we think only by the rehabilitation of the sensible world (compare Uexküll, the melody), not as a 'psychological fact' to reconstruct in positive terms, but as the visibility of the invisible. Compare Goldstein: the organism-milieu" (N, 303/238–39). As seen before with the notion of the melody, the body is not a thing independent of the environment but rather forms a unique melody as a whole. But now the body can be better appreciated for the activity *and* passivity implied earlier by the melody: the melody sings through the body because the body is perceptible. The body sings and is heard, just as it touches and is touched, sees and is seen, each implying both a unity with an environment as well as a fundamental divergence from it. The melody that sings through the body appealed to Merleau-Ponty, I believe, precisely because of the coupling of its activity–passivity (cf. VI, 183/139; 314/261;

318/265), a coupling that he expresses most forcefully in the case of the two hands touching. The reversibility or chiasm of the hands touching finds a similar expression in Uexküll's account of the melody when the animal both produces and is produced in its reciprocity with an environment.[23] The animal is neither subject nor object, but reciprocally producer–produced like a pure wake, a quiet force in the leaves of being. The melody, therefore, begins to approximate the flesh insofar as it extends and prolongs the body into the environmental world and is likewise incorporated by it. Although Merleau-Ponty does not resume an account of the melody here, he often continues to appeal to Uexküll's *Umwelt* as a particularly favorable manner of considering a body's immersion into the sensible; it sinks into and is enveloped by the flesh of the world because it is itself flesh. My body, he writes, is made of others' corporeality, each giving expression to a new kind of symbiotic ontology revealed through behavior.

In Merleau-Ponty's descriptions of space, the world does not play out in front of oneself as if the eyes were there to behold a two-dimensional spectacle. The body is enveloped by space just as much as other things are. This entails that one neither perceives one's own back and what happens 'behind' oneself nor does one perceive the other side of things in the world. In part, this phenomenological approach abstains from a greedy epistemology that seeks to grasp things as such. Even my own self is undercut by the nonpositivity of an invisible visibility. And yet, within the sensible world in which the body is immersed, one nevertheless encounters things. But how? And what are these things?

> What is there? First—visible or sensible being, things with their hidden "sides." Among the things are bodies which also have their hidden sides, their "other side," their being for the living (that is, not in that it is a consciousness, but in that it has an *Umwelt*). That is not constituted by our thought, but lived as a variant of our corporeity, that is, as the appearance of behaviors in the field of our behavior. (N, 338/271)

The *Umwelt* reveals that one not only approaches other things behaviorally—that is, not simply as a conscious perception—but that one also approaches oneself in precisely the same manner. To better illustrate this, Merleau-Ponty offers an enticing example of the perception of a cube. Just as I cannot perceive the hidden side of the cube, neither do I see myself seeing it because the reversibility, while always imminent, is never completely realized (VI, 193/147). There is absence and invisibility within the visible itself—not as the invisible part of the hidden cube, but as the invisibility within the being of the visible itself—and this is precisely the bodily dimension in which Merleau-Ponty roots the rehabilitation of the sensible. The other—the cube, another organism, or any other thing—completes my own unity insofar as both participate in the flesh of the world. I do not see myself seeing, but the other "closes the circuit and completes my own being-seen" (256/202). I come to be myself because of the reversibility provided by others. But it is a "blurring" of distinctions. The living being becomes itself in the thickness of this union, though it is a union of difference, separation, and invisibility.

The thickness of topological space also leads Merleau-Ponty to appreciate its circularity as opposed to a depiction of linear planes and layers. The layers and folds will still be an important aspect of his thought, specifically with respect to the 'archeological' image of digging into the hollows of being and getting beneath the accumulation of cultural sediment. But in terms of the structures supported by the dimensionality of being, circles and rings prove to be a more rewarding way of characterizing the reversibility of living things. The circularity is what is enacted within the reversibility of the flesh, where each body is circled because it extends into and is enveloped by the sensible. To this end, the body constitutes a "nexus" within the visible: "there is a relation of the visible with itself that traverses me and constitutes me as seer, this circle which I do not form, which forms me, this coiling over of the visible upon the visible, can traverse, animate other bodies as well as my own" (VI, 185/140). Again we glimpse a passive connotation of the body, where the body seems to simply endure being circled within the traversing of the visible. The circle forms the body, but the body is not wholly innocent either. Between activity and passivity, the circling appears like the melody, and finds a better expression as a "neutrality": "the animal body defined by the *Umwelt*, i.e., as aspects of the world cut up and organized by movements. Neutral between interior and exterior of the body. Intertwining or movement and perception. Neutral between centrifugal and centripetal" (N, 283/221). At the beginning of each of the first three "sketches" to Merleau-Ponty's third course, he opens with such a remark concerning the animal body. The human body, though different, is similarly characterized as receiving its *Umwelt* due to the relational circularity that is both active and passive, hence the appropriateness of neutrality. And if we're not mistaken, isn't neutrality similarly advocated in *The Structure of Behavior* to best exemplify the importance of behavior? Only this time, neutrality addresses the passive–active connotation of the animal body, which has an ontological significance that was lacking in the earlier assessment of behavior as simply being impartial toward the existing sciences.

But it is not only the case that there is a circularity in the *Umwelt*'s reciprocity with the body. Such an image connotes a circular process flowing back and forth between *Umwelt* and body, or between the sensible and the sentient. It is also the case, however, that we are led to envision the body itself—and the *Umwelt* itself—as spherical, much like we saw in the case of Heidegger's writings on the "encircling rings." For example, consider the following:

> My body as a visible thing is contained within the full spectacle. But my seeing body subtends this visible body, and all the visibles with it. There is reciprocal insertion and intertwining of one in the other. Or rather, if, as once again we must, we eschew the thinking by planes and perspectives, there are two circles, or two vortexes, or two spheres, concentric when I live naively, and as soon as I question myself, the one slightly decentered with respect to the other. (VI, 182/138)[24]

Initially this reads rather peculiarly, for what does he mean by these two circles being concentric or decentered? It isn't immediately obvious. But on further reading, Merleau-Ponty is depicting a circular model such that the body and its *Umwelt*, each a circle (or vortex, sphere) of its own, together form a concentric whole whereby the two circles invisibly overlap one another as one. It is only when reflective thought intrudes, jarring oneself out of 'naïve' being, that the two circles decenter and show each other as separate, as if having been thrown into a world of the present at hand. Again, it is a peculiar image, though no less a forceful one.

What are we to make of this circling? To better situate these circles, it would be advantageous to recall a passage that we have already considered, in which the animal is produced by the production of a milieu. We noticed the neutral reciprocity involved here, but there is also the interanimality that he expresses in the crossing of two *Umwelten*, the two rings of finality, such as that between the rat and the viper (N, 227/173). Though this is just one example, it is illustrative of the multiplication of circles that coincide in the structure of being. The circle that each animal forms with its *Umwelt*—the concentric circle lived naively—overlaps with the rings of other living beings, all together intersecting and crossing with each other, each a chiasm with the other. This does not mean that each animal is "open" to all others—as critiqued by Heidegger in *The Fundamental Concepts of Metaphysics*—but that, *as* a living body, each animal lives within the sensible and therefore is engaged in its own dimensional relation to the flesh of the world. Does this mean that all beings together encompass a synthetic world? Not for Merleau-Ponty, who notes that these multiple chiasms form "one" only "in the sense of *Uebertragung*, encroachment, radiation of being" (VI, 315/261), which is the sensible itself.

The circular corporeal schema owes at least a little bit to Uexküll's theory of the *Umwelt*. At the beginning of the first sketch, Merleau-Ponty draws a direct connection between the body's relation to the *Umwelt* and the necessity to resume an interpretation of this reversible relation with respect to the flesh (N, 270–71/209). After the analysis we have presently undertaken, we are now in a position to better understand how such structural relations compose a view of the animal as a 'self.' The new ontology is founded in the circles that extend and subtend one organ (e.g., a finger), one organism (e.g., an ape, a human), and two or more organisms (e.g., a rat and vipers, myself and other people). In every case, the unity of the organism is always one of simultaneous divergence due to its reversible relation with an *Umwelt*, but such that it is also always immersed within the flesh:

> The sensoriality, its SICH-*bewegen* and its SICH-*wahrnehmen*, its coming to *self*—A self that has an environment, that is the reverse of this environment. In going into the details of the analysis, one would see that the essential is the *reflected in offset*, where the touching is always *on the verge* of apprehending itself as tangible, misses its grasp, and completes it only in a *there is*— . . . The flesh is this whole cycle and not only the inherence in a spatio-temporally individuated this. (VI, 313/260)

The being of the flesh is what completes each living being. Another way of putting this is that each living being goes out into its environment, is 'one' with its *Umwelt*, but only finds itself in the sensible, fleshly being of "there is," like a melody unfurling itself through nature.

INTERANIMALITY

It has not been my intention to suggest that Merleau-Ponty owes an insurmountable debt of gratitude to Uexküll's formulation of the *Umwelt*. This is no more the case than it was for Heidegger. However, given the extent that Uexküll's *Umwelt* figures into Merleau-Ponty's thought, it leads one to suspect that there is more than a hint of interest present. By this I mean to suggest that the *Umwelt* provides another source, or another layer if you will, to Merleau-Ponty's development of natural being. Far from being just any old biological theory, Uexküll's *Umwelt* aids Merleau-Ponty in considering the structural relation between organism and environment in such a way that he surpasses the subject-object distinction and instead posits a sensible layer uniting all of life in the flesh of the world.

By broaching his phenomenological observations with certain developments in modern biology, Merleau-Ponty helps pave the way for what might be learned from an interaction between philosophy and the sciences. Drawing from the biology of Uexküll and others, we discover how the *Umwelt* provides a source for his new ontology of nature, where being reveals itself allusively in the leaves and folds between bodies. Being, in other words, arises in the intersection that the *Umwelt* is—"*Umwelt* (that is, the world + my body)" (N, 278/216)—because the body, as intercorporeal, is full of the world. More prominently, an articulation of the flesh of the world—as a sensible *there is*—reveals how brute being is a structure—he speaks of "the structure of being," "structural ontology"—that emerges in the simultaneous intertwining-chiasm effected by the living body. Such is the significance of the in-between, of "interbeing." As expressly seen in the opening pages of his third *Nature* course, Uexküll's *Umwelt* leads directly into a consideration of the moving body as the basis for his theory of the flesh. Merleau-Ponty's new ontology thus finds a seed germinating in Uexküll's theory of the *Umwelt*, something already recognizeable in *The Structure of Behavior* but not fully entertained until his *Nature* lectures.

Like Heidegger, Merleau-Ponty appeals to the *Umwelt* as carrying philosophical import, but whereas Heidegger uses the *Umwelt* as the basis for considering the ontological differences between animal environments and human world, Merleau-Ponty's interest is directed more to how its melodic undertone parallels a theory of nature overall. The body's cohesion with its milieu therefore occupies a greater place of distinction within Merleau-Ponty's thought insofar as nature shows itself as an ontological leaf of brute being. Unlike Heidegger, Merleau-Ponty is not as conceptually rigorous when it comes to defining and distinguishing between animal being and human being, between environments and worlds. But this is also part of his ontology's novelty: his investigations of nature allow a more immediate participation between humans and animals in the same source of life. The point is not to argue that being human is the same as being every other sort of animal, but that all manners of life partake in the whole of natural ontology. For instance, "We must say: Animality and human being are given only together, within a whole of Being that would have been visible ahead of time in the first animal had there been someone to read it. Now this visible and invisible Being, the sensible, our *Ineinander* in the sensible, with the animals, are permanent attestations, even though visible being is not the whole of Being, because it [Being] already has its other invisible side" (N, 338/271). Thus, even though Merleau-Ponty will claim that the human body presents another manner of corporeity than that of the animal (N, 269/208; 277/214), the human and animal are not separated by an ontological abyss.[25] One can see that Merleau-Ponty highlights realms of disparity, but, as shown in his final working note, he is more concerned with "a description of the man-animality *intertwining*" (VI, 328/274) than he is with highlighting differences between things. In the end, he is less interested in staking a difference between animals and humans than he is with revealing brute being in "one sole explosion of Being."

This means, however, that no single definition emerges of the organism. An organism comes to its self—*is* its self—in its sensible being. Instead of a redefinition of the organism as Merleau-Ponty envisions it, we are more inclined to find an account of "life." This elision may not be accidental, for "life" may in this sense be uncannily similar to what Heidegger called "process" and "motion" in place of "organism." For example, Merleau-Ponty writes in one long meandering sentence:

> Dissociate our idea of Being from that of the thing: life is not a separable thing, but an investment, a singular point, a hollow in Being, an invariant ontological relief, a transverse rather than longitudinal causality telescoping the other . . . the establishment of a level around which the divergences begin forming, a kind of being that functions like a vault, statistical being against the random, overcoming by encroachment, ambiguity of the part and the whole (against Driesch: The machine is not actually reaction of all its parts), thus being by attachment, that we cannot grasp apart, not bring it close (like a hard nucleus), refusal of all or nothing. (N, 302/238)

If this description does not help to clarify Merleau-Ponty's interpretation of life, neither will this: "Life = being by sketch or outline, that is, territories, regions = inherence in increasingly more precise places in a field of action or a radiation of being" (303/238). Or does it? He already used the phrase "field of action" in his second lecture course, and it arose, as it does here, in his initial reading of Uexküll. It is as though Merleau-Ponty, rather than pointing toward a specific definition of the organism, instead chooses to lean, as he does so in this third lecture course, on an understanding of life in general. It is not this entity or that being that is of interest (as if life could be "a separable thing"), but rather the ambiguous, allusive, soft tissue of flesh that forms a structural network within a certain "place" or "territory." The language he uses is intentionally ambiguous: place, field, territory, region, each becoming increasingly more precise but never fixed as such. It is not fixed, in part, because it depends on the moving body, behaviorally embedded in being, but also in part because his sketches on nature are becoming increasingly informed by theories of ontogenesis as he looks more and more closely at the origins of life in all of its glorious fields, folds, envelopments, divisions, encroachments, and mutations.

By emphasizing life, it is not that Merleau-Ponty wishes to abandon the concept of the organism. He doesn't abandon it and shows no strong sign that he intended to. But even though the organism has been present since his earliest publications, it is also fair to say that the concept emerges transformed and cloaked in a new vocabulary by the end of his life. Uexküll's *Umwelt* aids in this transformation, as it also will in the case of Deleuze's interest in Uexküll's biology. The unlikely adoption of Uexküll's thought within contemporary continental philosophy continues with Deleuze who, like Heidegger and Merleau-Ponty before him, is intrigued by the philosophical utility of this biologist for ontological considerations. But whereas Heidegger and Merleau-Ponty are more inclined to maintain the concept of the organism despite their reservations with its terminological history, Deleuze is not so hesitant. The organism is declared the enemy as he gives full way for a thinking of virtual intensities, affects, bodies, milieus, territories, rhythms and refrains, and for what might emerge within a poststructuralist interpretation of living beings.

CHAPTER 5

The Animal-Stalks-at-Five-O'Clock
Deleuze's Affection for Uexküll

PROBLEMATIC ORGANISMS

In the thought of Gilles Deleuze, the organism poses both a problem and a solution. This is both a good and bad thing, though not in a manner that we might initially suspect. Following the lead of Antonin Artaud, the organism is declared the enemy of the body and of life. "The organism is the enemy" (ATP, 196/158), writes Deleuze, in collaboration with Félix Guattari, in A Thousand Plateaus.[1] It is not Deleuze's only enemy, to be sure, but it certainly exemplifies a kind of conceptualization of life that requires further probing. It is a curious call to arms and one that has nothing to do with a dislike of organisms or animals. It is nothing of the sort. Rather, it is more an issue of "going beyond the organism" (FB, 47/39), of penetrating past the phenomenological interest in the "lived body" and "being-in-the-world," in order to discover the ontological processes that create what we are accustomed to calling the "organism." The organism is the enemy.

We might begin by understanding this view of the organism by framing it as a solution to a problem. Contrary to common sense, the organism is partially the enemy because it presents a solution: "An organism is nothing if not the solution to a problem" (DR, 272/211). The reason why this might run counter to common sense is that a solution, in this instance, is not sought after. Whereas we are accustomed to looking for solutions to everyday problems—and often work hard to achieve goal-oriented answers—it is the case that, in the scenario presented by Deleuze, solutions are not the aim. We might even go so far as to say that solutions are an altogether bad thing. In this respect, he remarks in Difference and Repetition, a book that he claims to be the first wherein he tries "to 'do philosophy'" for himself (xv), that solutions and responses are an inadequate justification to stop thinking. A solution, in other words, fixes something in its place, reifies it, and conceals the problems that are always immanently there and in need of being questioned. Just as he will later state that learning is an infinite task and that problems are the true ontological import of questioning, solutions tend to abruptly replace the movement of problems with the rigidity and stasis of a true cover-up (216/166). In a passage that exhibits Nietzschean flair, Deleuze writes: "There are no ultimate or original responses or solutions, there are only problem-questions, in the guise of a mask behind every mask and a displacement behind every place" (142/107).

In contrast, problems and questions are what frame the ontological agenda of modern thought. "It must be remembered," Deleuze writes, "to what extent modern thought and the renaissance of ontology is based upon the question-problem complex" (251–52/195). But it is not just any question-problem; it is the question of being. Alain Badiou, in his book on Deleuze, puts it in the following manner: "Our epoch can be said to have been stamped and signed, in philosophy, by the return of the question of Being" (19). Insofar as this is the case, Badiou concludes that this is "why it is dominated by Heidegger."[2] Indeed, in as much as Heidegger reminds us that 'today' this question continues to remain forgotten, it is the task of fundamental ontology to renew the question of being. Heidegger's *Seinsfrage*, the question of being, commences his ontological analysis of human Dasein that will eventually sweep animal life into the fold. Deleuze, in part, upholds this task of challenging the meaning of being as one of a ceaseless and immanent questioning, and he does so in conversation with Heidegger as well as Merleau-Ponty, whom he acknowledges for the interrogative stance of *The Visible and the Invisible*.[3]

But Deleuze's project is not that of Heidegger's or Merleau-Ponty's. His ontology of the organism asserts itself in a direction by which the question of being takes on new sense. If the organism is a solution, it is at once a solution to its own problem. What is the problem? Later, it will be framed as follows: "The problem of the organism—*how to 'make' the body an organism*—is once again a problem of articulation, or the articulatory relation" (ATP, 55–56/41). Rather than going in circles, as it may appear that we are doing in going back and forth between problems and solutions, Deleuze is actually describing two simultaneous accounts of this living thing that we call 'organism.' On the one hand, we can call any and all living beings an 'organism': I'm an organism, a cat is an organism, you're an organism, that orchid over there is an organism. This offers a certain solution in that it accounts for the actuality of *this* or *that* living being. And yet, it presupposes many—no, a multiplicity, infinite—factors that go into the actualization of any being we call an organism. There is therefore also a problem—namely, how to describe the factors that contribute to the composition of living things. The problem of the organism is, from this perspective, the very question of Deleuze's ontology; in asking "how to 'make' the body an organism," Deleuze is asking us not to remain complacent with understanding the organism as a contained, stable, and self-same entity, since different factors coalesce into the making of the continual becoming of this thing we identify as an organism. In other words, rather than remaining satisfied with describing the organism as a certain substance of this or that type, with these or those qualities, extended in space and time, Deleuze is instead asking what goes into the genesis of this living process. What makes the body? What can a body do? What relations compose this individual? How does the body articulate itself? This is the problematic posed by the organism. It is both solution and problem.

In presenting this approach to the organism at the outset, I hope to suggest that the problem of the organism is indicative of Deleuze's ontology as a whole. Indeed, the organism is even noted as an "example chosen almost at random" (DR, 238/184), almost, that is, as a domain in which we might observe the application of Deleuze's ideas. Whether explicitly or not, living beings are present throughout the entirety of his thought and provide a particularly rich source for situating the problematic posed by an understanding of life that rests not on the stasis of being, but on a process of *becoming* in being. By noting this connection between Deleuze's ontology and the nature of organisms, my wish is to contribute to an increasing interest in how continental philosophy, as exemplified here by Deleuze, has engaged with advances made in the life sciences and natural sciences. In many respects, Deleuze leads the way in this regard, insofar as his thought incorporates theories from such diverse fields as quantum mechanics, thermodynamics, embryology, symbiogenesis, complexity theory, differential calculus, and so on. His attempt to modify contemporary ontology with a twentieth century picture of the physical, chemical, and biological world—that is, he offers a philosophical grounding for the revolutionary developments in modern science—has even led some commentators, such as Mark Bonta and John Protevi, to suggest that it may not be too much of a stretch to say that Deleuze is the Kant of our time.[4] Just as Kant reconceived philosophical thought around the tenets of classical science (Aristotelian time, Euclidean space, Newtonian physics), it is their contention that Deleuze does the same with "twisted time," "fragmented space," and "far-from-equilibrium thermodynamics." In a similar vein, Miguel de Beistegui has recently argued that, in light of the end of metaphysics that has dominated twentieth century thought, "it is contemporary natural science that constitutes the consummation of metaphysics."[5] More specifically, it is in Deleuze's philosophy that de Beistegui discovers "an attempt to reinvent philosophy in the face of modern science, yet in such a way that philosophy does not become passive and merely derivative in the process" (222).

Deleuze has therefore become a dominant source for rethinking the ontological basis of the world in which we live, and this has come primarily in relation to developments in biology and physics. Within the last decade alone, there has been a concerted effort to draw attention to what Keith Ansell Pearson has called a long neglected tradition in modern thought, namely, biophilosophy.[6] Along with Pearson and de Beistegui, Manuel de Landa is another who has emphasized the confluence of Deleuzian studies and science, with the publication of his influential *Intensive Science and Virtual Philosophy*. I mention these publications because, on the one hand, they have had a decisive influence in my reading of Deleuze's ontology, and, on the other hand, because I hope to add to this "neglected tradition" with my own modest reading of Deleuze. In order to do so, and in keeping with the outline of my overall project, I will keep my reading more or less confined to looking at how the theoretical biology of Jakob von Uexküll implants itself within Deleuze's writings. Whereas these other studies have engaged in a much more comprehensive interpretation of Deleuze's appropriation of contemporary science, I intend to keep my attention more squarely on how and why Uexküll's thought operates in parallel with the problematic organisms, milieus, territories, refrains, and animal becomings that populate this new ontological development. As was the case with Heidegger and Merleau-Ponty, Uexküll's thought plays an important role in the articulation of Deleuze's own ontological vision. Along the way, this reading will entail a better look at precisely why the organism is considered a 'solution' in relation to its more 'problematic' ontological status.

UEXKÜLL'S ETHOLOGY OF AFFECTS

It is perhaps unsurprising that Uexküll appears infrequently and sporadically in Deleuze's writings. Such is his style. In the early ontological works, such as *Difference and Repetition* (published in 1968) and *The Logic of Sense* (1969), one finds Deleuze discussing biologists such as Baer, Darwin, Geoffroy Saint-Hilaire, and Cuvier, but there is no mention of Uexküll as yet. And while there is a hint that Deleuze begins to approach Uexküll's thought within some of his lectures that he delivered in the early 1970s at Vincennes (e.g., in his occasional usage of the "tick" to describe the univocity of being), there is again no sustained, or, at any rate, direct reflection on how or why Uexküll may be advantageous to his thought.[7] To the best of my knowledge, an explicit engagement only occurs as early as 1978 with the publication of a very short, though no less important, essay entitled "Spinoza and Us."[8]

From this point on, Uexküll will appear like brief flashes in *A Thousand Plateaus* (1980), as well as in Deleuze and Guattari's final collaborative work, *What Is Philosophy?* (1991).

It is not my intention to attribute to Uexküll an inappropriate degree of influence on Deleuze. There is not a 'discovery' of Uexküll that changes the course of Deleuze's thought. Nor does he necessarily prove to be a missing link that had previously escaped Deleuze. Rather, Uexküll would be better served if we considered him as an additional dimension to the rhizomatic composition of Deleuze's thought, entering here and there as a heterogeneous relation and offshoot. Not essential to Deleuze, but not insignificant either. In this role, Uexküll can be appreciated as a relative curiosity, all the while appearing in a manner that evinces his broader appeal. For one, Uexküll proves insightful in his capacity as "one of the main founders of ethology," a discipline traditionally associated with the study of animal behavior but which Deleuze reinterprets in his conjunctive reading of Uexküll with Spinoza. Ethology, according to Deleuze, is a study of affects: "Such studies as this, which define bodies, animals, or humans by the affects that they are capable of, founded what is today called *ethology*" (SPP, 167–68/125). What I wish to argue here is that ethology is not merely a pet project for Deleuze, but rather speaks more pervasively to his general ontology insofar as it is concerned, at least in part, with a study of affects. Hence, Uexküll's appeal as one of the contemporary founders of ethology. The essay, "Spinoza and Us," wherein Uexküll stands shoulder to shoulder with Spinoza, incorporates many of the key ideas raised in Deleuze's ontology, and does so in an uncharacteristically clear manner (many of the ideas and examples raised here will be later recycled throughout Deleuze and Guattari's *A Thousand Plateaus*). Exploring this essay will provide us not only with an understanding of how Uexküll is officially introduced into Deleuze's project; it will also include a nice introduction to Deleuze's overall ontology. This is why, in my opinion, Uexküll proves to be such a fascinating figure, since "Spinoza and Us" revolves around him just as much as it does around Spinoza. "Uexküll . . . is a Spinozist" (169/126), Deleuze writes, and both are ethologists of a new order.

Ethology clearly means something different here than it has elsewhere. What, then, does Deleuze mean by claiming that ethology is the study of affects? One might begin, as he does, by offering an example, such as the example of the tick that Uexküll describes in his *A Stroll Through the Environments of Animals and Humans*. As we have already seen, this example of the tick has received plenty of mileage, probably due to its relative simplicity and directness. Yet its simplicity belies the variety of different interpretations that it produces. Deleuze offers us his reading: "[Uexküll] will define this animal by three affects: the first has to do with light (climb to the top of the branch); the second is olfactory (let yourself fall onto the mammal that passes beneath the branch); and the third is thermal (seek the area without fur, the warmest spot). A world with only three affects, in the midst of all that goes on in the immense forest" (167/124). What Deleuze recognizes in Uexküll's example is the attention paid not so much to the animal itself, but to what this animal can do. At issue, therefore, is how the tick relates to its surroundings, where the emphasis is neither on the tick (its species, its color, whether it has four or six legs, etc.) nor on the environment (this or that mammal, a tree, a bird, etc.), but on the "affective" relation itself. Here Deleuze departs from a more literal interpretation of Uexküll, where the various signs in the environment (e.g., the mammal) play a more significant role in Uexküll's original telling of this example (SAM, 7). But Deleuze is especially perceptive in highlighting how Uexküll counts three affects: light, scent, heat. These three affects alone define the tick. "You will define an animal, or a human being," Deleuze informs us, "not by its form, its organs, and its functions, and not as a subject either; you will define it by the affects of which it is capable" (SPP, 166/124). Everything outside of these three affects is a matter of indifference to the tick, as should be the case in our own look at animal worlds. Our cue, in other words, ought to follow from the affects of animals.

Counting affects is therefore the task of ethology. There are a few things to note here. First, counting does not mean establishing quantitative difference. A world with only three affects is neither more nor less than another world. Whether it even counts as a 'world' is another question entirely and one that will be addressed later; for the moment, let me just note that Deleuze fails to adequately conceptualize "world" but that he does so for good philosophical reasons. As for the three affects, they constitute an 'ordinal' series (first, second, third) as opposed to a numerical value based on quantitative cardinal numbers (one, two, three). Manuel de Landa explains how this mathematical distinction is an important one in Deleuze's thought, particularly in how a cardinal series may be added or divided into increasingly more or less metric numbers (1.1, 1.2, ... 2, 2.1 ...), whereas "an ordinal series demands only certain asymmetrical relations between abstract elements, relations like that of *being in between* two other elements."[9] In the case of the tick, we observe how the three affects are described ordinally, such that the life of the tick is composed of a procession through this sequential order, irrespective of what happens in between. The tick might live in a dormant state for many years between the first and second affect; between affects, literally nothing affects it. Each affect instantiates a new 'becoming' in the tick's life. Second, and along the same line, the affects themselves are of an ordinal nature, by which I mean that they are not states that can be counted and divided without changing their nature altogether. Solar light, olfactory scent, thermal temperature: each constitutes a dimension immanent to metric space, beyond what Deleuze calls the "extensive." Extensive properties refer to the classical states of measurable Cartesian space, such as length, breadth, and volume. In contrast, the affects denote "intensities," a difficult concept in Deleuze's ontology and one to which we will have to return, but can characterize for the moment as "an implicated, enveloped, or 'embryonised' quantity" (DR, 305/237). To better clarify this numerical distinction, Deleuze draws heavily from both quantum mechanics and thermodynamics, and appeals to examples involving temperatures, speeds, dilutions, and other qualities: "when it is pointed out that a temperature is not composed of other temperatures, or a speed of other speeds, what is meant is that each temperature is already a difference, and that differences are not composed of differences of the same order but imply series of heterogeneous terms" (306/237).[10] The range of temperature, speeds, or light, are a matter of degree that cannot be added or divided without changing the nature of the thing. A common example in this case is the temperature of water that, at certain critical points, transforms its compositional 'nature,' such as from water into ice or gas. It hits a tipping point and becomes something different. Deleuze concludes that "an intensive quantity may be divided, but not without changing its nature." This description is applicable to all intensities, such as color, tastes, light, or even, as he expresses in his evaluation of Lucretius's philosophy, the warmth of the fire and the liquidity of water (LS, 320–21/277).

This means that unless the tick's sensory organs detect solar light, the precise scent emitted by the butyric acid of a mammal, or the exact temperature of mammalian blood, the tick will not be affected. A difference in photosensitivity, olfactory scent, or thermal temperature would lead to no affective relation. Another way to describe this relation is that the tick can be affected by a 'dummy' or surrogate that does not have blood but a similar liquid heated to the temperature of blood. The affective relation is therefore not between the tick and the mammal, but between a sensory organ and light, scent, or heat. To the extent that an affect changes, the entire scenario changes as well, including how we understand this relation. Each affect is both ordinal and intensive.

Even after looking briefly at the example of the tick, however, it is still unclear what affects are. A clue can be found with Spinoza. If ethology is the study of affects, and if Uexküll counts affects, and if Uexküll is, in Deleuze's mind, a Spinozist, then we might do well to see how Spinoza defines affects. This is, after all, one of the primary sources for Deleuze's usage of this concept. Most of *Spinoza: Practical Philosophy* comprises a conceptual index of Spinoza's key terminology, which includes among them 'Affections, Affects' (SPP, 68/48). There are three ways that Deleuze interprets Spinoza's concept of affect: (1) affections (*affectio*) are modes of substances; (2) these modes also designate "the effects of other modes on them," their modifications; and (3) the modes or affections are transitions or passages of states called affects or feelings (*affectus*). One might summarize the terminology as follows: "The *affectio* (affection) refers to a state of the affected body and implies the presence of the affecting body, whereas the *affectus* (affect) refers to the passage from one state to another, taking into account the correlative variation of the affecting bodies" (69/49). In Deleuze's reading, affects are inseparable from their relations to bodies, where bodies are not considered in terms of their function or form, but in terms of how they can be affected, how they can undergo transitions, and, at bottom, how they define what a body can do. For example, in his first book on Spinoza, *Expressionism in Philosophy: Spinoza*, Deleuze explains that "a horse, a fish, a man, or even two men compared one with the other, do not have the same capacity to be affected: they are not affected by the same things, or not affected by the same things in the same way" (197/217). Each living thing entails a different set of relations to the environment at large. The capacity to affect and be affected is what constitutes the individuality of each particular thing. He continues: "A body's structure is the composition of its relation. What a body can do corresponds to the nature and limits of its capacity to be affected" (198/218).

While Deleuze adopts the concept of affect from Spinoza, these descriptions are more revealing in terms of what they say about his own image of bodies. Where affects refer, in Spinozist terms, to the passages and transitions between bodies (as modes, not substance), affects can be concisely defined, in Deleuzian terms, by observing that "affects are becomings" (ATP, 314/256). We will have plenty of time to discuss the importance of 'becoming,' but we can see here that affects, through their affinity with becomings, involve a passage or transition between states of a thing, as well as—and this cannot be thought separately from the preceding—a passage or transition between different bodies. In stating that ethology is a study of affects, Deleuze is therefore noting a convergence between differential relations that compose an individuated thing. Before one accounts for these relations—in other words, before one begins counting affects—one knows little about the thing (horse, tick, human, orchid . . .) in question. Deleuze and Guattari suggest as much in reference to the body: "We know nothing about a body until we know what it can do, in other words, what its affects are, how they can or cannot enter into composition with other affects, with the affects of another body, either to destroy that body or to be destroyed by it, either to exchange actions and passions with it or to join with it in composing a more powerful body" (314/257).

One can see how the concept of ethology has been radically reenvisioned here. Ethology has less to do with the study of behavior per se than it does with the continual becomings that compose different bodies. "In short," we're told, "the notion of behavior proves inadequate, too linear, in comparison with that of the assemblage" (410/333).[11] But if we are not to study readily identifiable properties pertaining to animal behavior, how do we count affects? In other words, what is it that is 'measured' when considering affective becomings? The example of the tick has already indicated the kind of relation that Deleuze is after. But we are only beginning to speak of the possibilities for how things affect and are affected—or, in other words, how things become. Another example is that of the differences between a workhorse and a racehorse or ox: despite taxonomical classifications, Deleuze suggests that the workhorse has more in common with the ox than it does with the racehorse because the workhorse and ox have affects in common (SPP, 167/124; ATP 314/257). Insofar as affects account for passages, movement, becomings, both within and between individuals, Deleuze will emphasize that it is the process of such modifications that require elucidation. "Ethology," therefore, "is first of all the study of the relations of speed and slowness, of the capacities for affecting and being affected that characterize each thing. For each thing these relations and capacities have an amplitude, thresholds (maximum and minimum), and variations or transformations that are peculiar to them" (SPP, 168/125). In the case of the workhorse, its "body" is closer to that of the ox than the racehorse. An ethological study will reveal this. For the tick, it is the degree of light, an olfactory scent, a thermal register. The tick, in other words, is the composition of these relations, and each tick will differ in its capacity to be affected and to affect according to these different degrees. In an Uexküll-inspired passage, Deleuze writes that "every point has its counterpoints: the plant and the rain, the spider and the fly. So an animal, a thing, is never separable from its relations with the world" (168/125). These counterpoints are not restricted to a binary coding, but indicative of the many more relations possible for the individuating process within each thing.

In a manner that parallels the earlier threefold definition of affects, Deleuze adds two more descriptions to the preceding redefinition of ethology. Not only must ethology study the affects of each thing (speeds and slowness, to affect and be affected), but also the circumstances that determine how and whether such relations may be successfully entered. Affects not only *are* a process of bodies, but they are so in different ways. It is always a process of differentiation. For example, Deleuze notes that affects do not simply affect a given thing, but that, as a becoming, they might threaten, strengthen, accelerate, increase, decrease, or even destroy the body. Affects have an effect on bodies that complicate the relations themselves. As a result, the affective becomings change the body in question. The third and final aspect of ethology is that it "studies the composition of relations or capacities between different things" (169/126). Part of Deleuze's ontology, as we will see, is that bodies can enter into greater and more powerful bodies, as well as weaker and destructive ones, through the composition of relations between different things. For example, a plant may have a certain relation with rain, but it will also have multiple relations, such as with the acidity of soil, the solar rays of the sun, other plants, birds, and so on. Through the composition of these relations, the plant may form a 'higher' individual, such as a symbiotic unit, a garden, forest, ecosystem, and so on which is itself its own body with its own speed and slowness, ability to affect and be affected. Similar examples may be found in an infinite number of ways, from human communities to coral reefs to labor forces to linguistic practices.

In stating that we know nothing about a body until we know what that body can do, in its capacity to affect and be affected, we are also led to reconsider what we mean when we speak of the body. It is surely not the phenomenological body of lived experience of which Deleuze speaks. The "lived body is still a paltry thing in comparison with a more profound and almost unlivable Power" (FB, 47/39), a power that is more akin to a rhythmic unity plunging into a chaos of relations. The body is not the lived body, then, and nor is it the scientific body in *res extensa*. Rather, "a body can be anything; it can be an animal, a body of sounds, a mind or an idea; it can be a linguistic corpus, a social body, a collectivity" (SPP, 171/127). Cloud, feather, rainbow, car, justice, spider-fly, a pack of rats. Each is a body. A body is the accumulation of relations that, at a certain state of equilibrium, is able to maintain itself without destroying itself as *this* consistency. However, as we have seen throughout the foregoing discussion, each and every thing is literally in a process of composition and decomposition. Every body, as affective, is a becoming. So how are we to speak about the individuality of any given body insofar as it is always in a relational state? It has already been suggested that one defines an animal or any other body "by the affects of which it is capable." Yet, even after the analysis on affects undertaken above, we have only begun to break the surface of Deleuze's ontology. What applies to animals may also be applied to life more generally: "The important thing is to understand life, each living individuality, not as a form, or a development of form, but as a complex relation between differential velocities, between deceleration and acceleration of particles. A composition of speeds and slownesses on a plane of immanence" (165–66/123). Indeed, ethology applies not only to individuated bodies like animals, nor just to organic life in general, but to everything, since "the plane of immanence, the plane of Nature that distributes affects, does not make any distinction at all between things that might be called natural and things that might be called artificial" (167/124). His ontology is universal, or, more appropriately, as we shall soon see, "univocal." To be able to address the actuality of a body—where a body can literally be *any* thing—we must explore further the components of Deleuze's ontology, wherein he speaks equally of micro- and macrophenomena, of the molecular and the molar, the intensive and the extensive, the virtual and the actual.

Throughout this discussion, it has been my hope that the problematic nature of organisms has started to emerge more clearly, even if indefinitely. An organism proposes a solution only at a level where one has stopped counting affects, where the body has been taken as a formal and fully organized self-consistency. At this point, however, its reification has concealed the affective relations that continually compose life. In other words, the actualization of the individual organism relies on a simultaneous and immanent set of "complex relations" that play out across what Deleuze here calls "the plane of immanence." The notion that bodies are made of differential velocities and the movement of particles has given rise to various claims that his ontology is an "ontological materialism" or a "realist ontology."[12] While Deleuze's philosophy cannot be confused with classical atomism—his essay "Lucretius and The Simulacrum" captures his closer proximity to a 'quantum' form of atomism (LS, 307–24/266–79; cf. DR, 238/184)—it does require that one pass beyond the surface, so to speak, of actual bodies. His ontology is therefore just as much a "quantum ontology," an "ontology of the virtual," and/or "an ontology of intensities," as it is a different kind of materialism.[13] "The body," Deleuze and Guattari write, "is now nothing more than a set of valves, locks, floodgates, bowls, or communicating vessels, each with a proper name" (ATP, 189/153). The body is a porous composition, continually transforming itself, becoming-other, even if imperceptibly. To borrow from Deleuze's reading of Lewis Carroll in *The Logic of Sense*, we must make ourselves like Alice and pass through the rabbit hole of the body to look more closely at what makes the intensive body. What must now occur, therefore, is that we dismantle the organism by "opening the body to connections that presuppose an entire assemblage, circuits, conjunctions, levels and thresholds, passages and distributions of intensity" (198/160). In order to further count affects, to fully engage in this ethological project, that is, we must proceed by unraveling the reified solution posed by the organism by revealing its problematic composition: how the body is 'made' an organism. With this, we pass to the body without organs.

THE BODY WITHOUT ORGANS, THE EMBYRONIC EGG, AND PREBIOTIC SOUP

As previously remarked, Deleuze's ontology is often tricky to track in its constantly emergent state, situated as it is between what he calls the "actual" and the "virtual." To complicate matters further, the actual, the virtual, and the intensive are all different ways of addressing the "univocity of being" that highlights Deleuze's ontology. In *Difference and Repetition*, Deleuze emphatically states that "there has only ever been one ontological proposition: Being is univocal" (DR, 52/35; cf. LS, 210–11/179–80). Most of his book, and indeed much of his early thought, is devoted to clarifying this very claim, as evinced by the conclusion's final words: "A single and same voice for the whole thousand-voiced multiple, a single and same Ocean for all the drops, a single clamor of Being for all beings: on condition that each being, each drop and each voice has reached the state of excess—in other words, the difference which displaces and disguises them and, in turning upon its mobile cusp, causes them to return" (389/304). This single voice is said to echo from Parmenides to Heidegger and to speak to all of being. Considering what we know of Heidegger's ontological difference, not to mention the ontological distinctions between ontic beings, this claim is sure to raise an eyebrow or two. Among his skeptics is Alain Badiou who doubts that Deleuze ever posits a reliable concept of multiplicity within his univocity. In other words, Badiou thinks that Deleuze is a little bit too convincing in his appeal to the One, to the extent that he does not posit a true theory of multiplicities (53, passim). Does the single clamour of being truly reverberate across all things and through all beings? Does this mean that every thing is ontologically equivalent? More to the point, are we to understand that all beings—from gods to humans to animals to plants to inanimate bodies—are ontologically indistinct? Is there any compliance between Deleuze's univocity of being and both Heidegger's and Merleau-Ponty's respective ontologies?

Deleuze draws his univocity of being from three primary sources: Duns Scotus, Spinoza, and Nietzsche.[14] Since Deleuze's readings of these three figures have received plenty of attention in secondary literature, I'm principally interested in looking at how the univocity of being functions with an orientation toward animal life.[15] One of the key components of Deleuze's account of univocal being is that, beginning with Scotus, it has been conceived as a direct alternative to an analogical approach to being. To briefly summarize analogical being, this ontological position speculates that being is said in many different ways, but that these different ways may all be related through analogy. The great chain of being that extends from God and angels to plants and rocks can be conceived analogically, which posits both a form of resemblance as well as difference between beings. For example, St. Thomas Aquinas held that we cannot know the essence of God other than by way of analogy, by appealing first to God's creatures in order to ascend, analogically, to an understanding of God (cf. LS, 210/179). But this analogical approach must be distinguished from both a 'univocal' and 'equivocal' sense. With the former, when we predicate that humans are wise and that God is wise, we use 'wise' analogically but by no means in the same sense. Yet, with this example, the word 'wise' cannot be purely equivocal either, for otherwise its usage would be trivial and utterly meaningless due to the lack of intrinsic similarity. Thus, according to analogical being, being can be said of both God and creatures, but it can be said neither in a univocal sense (they do not possess being in the same sense), nor in an equivocal sense (they both may be wise, albeit differently). In the end, "analogical predication is founded on resemblance."[16]

Resemblance, however, along with identity, analogy, and opposition, are not popular terms within Deleuze's apparatus.[17] The main problem that Deleuze has with analogical being is that it prioritizes resemblance and identity over difference itself. In other words, analogy, identity, similarity, opposition, and resemblance are grounded by the act of erasing the differences that make the terms distinct. As for equivocal being, it does not prove to be as big of a problem as analogy, since Deleuze applies univocity to the equivocal differences among beings: "Univocity signifies that being itself is univocal, while that of which it is said is equivocal: precisely the opposite of analogy" (DR, 388/304). Therefore it does not matter what univocity is said with respect to because being is univocal, no matter the entity. With analogy, there is a comparison between terms, a bridging and dissolving of difference, as opposed to an attempt to look at the differences that compose the terms themselves. Deleuze summarizes as follows: "It is henceforth inevitable that analogy falls into an unresolvable difficulty: it must essentially relate being to particular existents, but at the same time it cannot say what constitutes their individuality" (56/38). What constitutes the individuation of entities is the play of difference itself. Therefore, "resemblance, identity, analogy and opposition can no longer be considered anything but effects, the products of a primary difference or a primary system of differences" (154/117).

This brings us back to univocal being. By contrast to analogical being, Deleuze's philosophy emphasizes that being *can* be said in a single and same sense. In saying so, he is not merely fanciful when he invokes a single clamor of being for all beings. But what does this single clamor mean? Above all, it means that being is difference itself. "Being is said in a single and same sense of everything of which it is said, but that of which it is said differs: it is said of difference itself" (53/36). Being may be said to be univocal insofar as it applies to difference, not just any specific or generic difference, but a system of individuating differences. It is not necessarily a comparison between this or that being that ultimately concerns Deleuze, for, in doing so, one inevitably comes around to reasserting terms of analogy (e.g., categorical concepts) when comparing actual things. One discovers, therefore, in a lecture course that Deleuze delivered in the mid-1970s, that being can be said in the same way of such seemingly disparate beings as, for example, God and ticks. "In a certain manner," Deleuze explains, univocal being "means that the tick is God; there is no difference of category, there is no difference of substance, there is no difference of form. It becomes a mad thought."[18] This interpretation of univocity is Deleuze's own, and it indeed points toward a mad thought. One can see here that Deleuze takes equivocity to the extreme to better demonstrate his point: univocity applies to *all* of being, regardless of category, form, or substance, and it is equivocal toward that which it is said, whether it is said of God or ticks, rocks or clouds, angels or humans. Determinations of this sort are not what Deleuze finds to be the most engaging problems. Nor is it the case that, in noting that being can be said in the same way of both God and ticks, they are somehow *one* and the *same* being. We have already observed how different bodies are affected in different ways; indeed, the capacity to be affected in different ways is what allows for the differentiation of beings.

Rather, it is the play of differences that compose and individuate beings that is of particular interest. As mentioned earlier, this requires that we must look at individuating factors: "Univocity of being, in so far as it is immediately related to difference, demands that we show how individuating difference precedes generic, specific and even individual differences within being; how a prior field of individuation within being conditions at once the determination of species of forms, the determination of parts and their individual variations" (DR, 56–57/38). The emphasis placed on univocity directs our attention to the movement and genesis implicated within all of being; no matter what the body, being is said in the same sense because it is said of the play of difference within each individuation. If there is substance, then this substance can only be attributed to a "multiplicity"; for Deleuze, multiplicity "is the true substantive, substance itself" (236/182). Every body, all of being, therefore, must be investigated in terms of the emergent properties that coalesce in the genetic structure of each body's respective state. The variety of multiplicity—difference itself—is all that there is. One manner of speaking of this coalescence is as an "unnatural participation": "The rat and the man are in no way the same thing, but Being expresses them both in a single meaning in a language that is no longer that of words, in a matter that is no longer that of forms, in an affectability that is no longer that of subjects" (ATP, 315/258). The question then becomes one of "mapping" the ontological dimension of this univocity, which can only be described as one of difference. If everything is multiplicity, and yet univocal in this difference, then we must look more closely at the "onto-hetero-genesis" implicated within the becoming of divergent entities.[19] In order to do so, we will need to look more closely at the "intensive" and the "virtual" dimensions of Deleuze's ontology.

In the conclusion to *Difference and Repetition*, Deleuze remarks on the need to follow a path back from extended, 'actual' entities, like organisms, to another ontological dimension that contributes to the composition of their individuated states (DR, 360–61/282; cf. WP, 133/140). This return or path back has nothing to do with a regression, nor does it involve a preceding time or space. Rather, it has more to do with unfolding the plane across which various multiplicities take shape and emerge as more or less stable strata. The intensive and the extensive, the virtual and the actual, the molecular and the molar, each are manners of addressing the immanence entailed within various compositions. For example, the organism is not simply an extended entity that can be defined as a substance with various qualities, nor can it be adequately characterized as an individual belonging to this or that species. The problem with each of these determinations—as we have already observed with Heidegger—is that they continue to rely on the organism as a static or present-at-hand object. If there is a correlation between Heidegger and Deleuze's ontology, one area would be in terms of how both are critical of a static portrayal of being. To this end, they both speak to an ontological difference.[20] With Heidegger, on the one hand, there is a difference between being (*Sein*) and beings (*Seienden*), between ontology and the ontic. Within ontology, there is also the issue of fundamental ontology, which pertains to the particular manner of human existence for which being is said to be at issue. With Deleuze, on the other hand, there is a difference between the virtual and the actual, and, within this difference, there is a continual process of differentiation leading from one to the other. Despite this similarity, however, I think it would be a mistake to see the two as paralleling one another, as if Deleuze's virtual mirrors Heidegger's being (*Sein*), and actual extended things mirror beings (*Seienden*). The problem is not so much with the former as it is with the latter. Deleuze's actual is roughly equivalent with Heidegger's ontic beings, but the same does not apply to the virtual and being. Some have suggested this equivalence, but I simply do not find the virtual as approximating Heidegger's account of being.[21] Just as Heidegger was impatient with previous systems of thought that dwelled too much on the objectivity of things, and an ontic depiction of human existence, Deleuze is similarly impatient with the reification of living and nonliving processes. Beings—whether we are speaking of human beings or rocks—are not simply given in an unchanging, essentialist manner. With Heidegger, things are ready-to-hand within a horizon of preunderstanding.

In contrast, Deleuze asserts that each being is a unique composition that is continually in the process of becoming, and becoming at different levels, through different relations, and at different speeds and slownesses. If we consider Heidegger's "ready-to-hand" as a mode of "coming-to-presence," as Heidegger does, the proximity might seem greater, at least at first glance. At stake in Deleuze's ontology, then, is a careful account of the differences that underlie our view of what is actual. Organisms, for their part, must again become embryos, larval selves, though not in any developmental sense. This requires further explanation.

Deleuze contends that "we are never fixed at a moment or in a given state but always fixed by an Idea [i.e., a multiplicity: "Ideas are multiplicities: every idea is a multiplicity or a variety" (DR, 236/182)] as though in the glimmer of a look, always fixed in a movement that is under way" (283/219). Whether spatially or temporally, everything is literally in movement and already under way. Mountains that stand impressively on the horizon of our view are but fluid matters across the scope of geological and glacial time. The eighteen years that a tick might wait for a mammal to cross its path are nothing in comparison to the eighteen years a human might wait for a comparable event. Likewise, the flow of genes across some species and not across others, the rise and fall of various cities and civilizations, the appearance across the night sky of a comet that arrives every seventy-six human years, are all examples of the transformative nature pertaining to all things, whether living or nonliving, whether of single entities or of populations.[22] Underlying all of this movement is the process of differentiation, however great or minute it might be. In part, Deleuze problematizes the anthropocentric view that is often entailed when considering the question of being. For this reason, we are led to the dimension of multiplicities that compose beings, rather than taking things as though they were already fully formed, static, and isolated entities.

Deleuze's ontology is therefore that of a process that considers the moving, sliding, passing, shifting, and changing landscape that goes into the composition of something (ATP, 72/55). In the case of the organism, the interest here is in what makes this body and what this body can do. How does this living being emerge as *this* body in particular? What affects is it capable of creating and entering? Such questions could be answered biologically, socially, economically, politically, historically, psychoanalytically, and so on, but this still wouldn't get to the heart of the matter. They wouldn't speak to the ontology of the organism. And yet, it isn't even the organism that is of interest here. The organism, remember, is the enemy. Rather, Deleuze's focus is in the coagulating process of this specific entity we call an organism. There is a hidden movement of individuation behind every actual being: "Beneath the actual qualities and extensities, species and parts, there are spatio-temporal dynamisms. These are the actualising, differenciating agencies. They must be surveyed in every domain, even though they are ordinarily hidden by the constituted qualities and extensities" (DR, 276/214). Such is the work that Deleuze's thought brings to the fore. What produces and creates a body? In other words, one of the central questions at stake is how actualization happens.

Deleuze's emphasis on genetic—as opposed to static—being relies on the process moving between the virtual and the actual. The virtual is a real, nonlinear continuum of differentiation that creates actual things through intensification without in any way causing or resembling the actual product. Boundas explains the process as follows: "The differential quantity of forces is called 'intensity' and intensities—or better, intensifications—are the real subjects of processes. But they are not subjects in any ordinary sense since intensities are not entities. Being responsible for the genesis of entities, they are virtual yet real events, whose mode of existence is to actualize themselves in states of affairs [extended things]."[23] Intensities are not actual subjects capable of causing other actualities, which is why Deleuze refers to the virtual as a "quasi-cause" (LS, 149/124) or a "dark precursor" (DR, 156/119). Virtual intensities produce the actual state of affairs, and coincide immanently within the becoming of every thing, without, however, being identifiable with extended being. In a way, the virtual is like a constant embryological state of becoming, albeit with a peculiar Deleuzean twist.

It is not by coincidence that many of Deleuze's most telling ontological descriptions emerge in collaboration with his descriptive accounts of biology. Continuing from the preceding point, embryology is briefly discussed in order to exemplify at least three points in unearthing these hidden virtual processes. "Embryology," Deleuze writes, "shows that the division of an egg into parts is secondary in relation to more significant morphogenetic movements: the augmentation of free surfaces, stretching of cellular layers, invagination by folding, regional displacement of groups. A whole kinematics of the egg appears, which implies a dynamic" (DR, 277/214; cf. FB, 47-48/39). The first point to be noted is that there is a movement away from descriptions that rely on the terminology of whole-parts (a movement that we observed in Merleau-Ponty's discussions of animal life) toward a new terminology of virtual-actual. The embryonic egg is not a whole composed of different parts, but is instead the actualizing of a virtual continuum. To this end: "Rather than going from more to less general, determination progresses from virtual to actual in accordance with the primary factors of actualisation" (DR, 277/214). It should not be overlooked that this account of embryology derives from Deleuze's critical appraisal of Baer's biology, whose thought, including the development 'from more to less general,' we briefly treated in the chapter on Uexküll. This encounter will become more clear in a moment. The second aspect follows from the first; the embryo develops due to a variety of different processes, each of which may have any number of results, whether accidental, ineffectual, transformative, delaying, crippling, and so on. The egg is not a whole composed of parts, but an unfolding process of individuating factors that bear on this egg, and this egg only. To this end: "Types of egg are therefore distinguished by the orientations, the axes of development, the differential speeds and rhythms which are the primary factors in the actualisation of a structure and create a space and a time peculiar to that which is actualised" (277/214). Every embryo, even those belonging to the same species, develops at different rates and speeds, varies according to environmental factors, undergoes cellular displacements that lead to any number of significant changes and repercussions for that being. A third cue from this interpretation of embryology is more of an implicit point—namely, that Deleuze is not offering a traditional developmentalist view of life. He is not arguing that an organism individuates according to a specific developmental plan, as though unfolding according to a strict and regimented schedule. There is no mechanist or vitalist appeal made here, no causal determination or teleological force. There are no progressive stages through which the egg must pass to become what it is predestined to become. Rather, as he will later note in connection with Darwin's inauguration of individual difference, "it is a question of knowing under what conditions small, unconnected or free-floating differences become appreciable, connected and fixed differences" (319/248).[24] At issue is this coagulating process of condensation that creates a 'fixed' and 'stable' multiplicity out of the flow of a differentiating field. In a similar vein, Manuel De Landa has also noted that "our organic bodies are, in this sense, nothing but temporary coagulations in these flows."[25] The embryo must be understood as the continual actualization of innumerable factors playing out on a virtual continuum.

In fact, the embryonic egg proves to be a particularly helpful introduction to what Deleuze is doing in his ontology of virtual intensities. Without simplifying the highly complex thought that underlies Deleuze's concepts, it is the case that he is concerned with how things are actualized from a preindividual plane of immanence. It just so happens that the egg becomes a peculiar demonstration of this plane of immanence as well as the unfolding or envelopment into a state of further actualization. I emphasize the egg for a few reasons. I do so, first, because the egg seems to occupy a particular place in Deleuze's thought. It enters in *The Logic of Sense*, for instance, as a means to relate his readings of Lewis Carroll and the Stoics; here Humpty Dumpty is like a Stoic master, and philosophy is comparable to an egg.[26] Of greater importance for our investigations is how the egg is interpreted with respect to embryology and, more significantly still, how it is tied to his ontological system. On the one hand, the egg serves as a particular model for the "world": "In order to plumb the intensive depths or the *spatium* of an egg, the directions and distances, the dynamisms and dramas, the potentials and potentialities must be multiplied. The world is an egg. Moreover, the egg, in effect, provides us with a model for the order of reasons: (organic and species related) differentiation-individuation-dramatisation-differenciation" (DR, 323/251). Deleuze's distinction between "differentiation" and "differenciation" is an important one in his ontology inasmuch as it addresses the processes of becoming that occurs in the virtual and the actual, respectively. On this distinction, Deleuze writes: "We call the determination of the virtual content of an Idea differentiation; we call the actualization of that virtuality into species and distinguished parts differenciation" (267/207). Again: "The *t* and the *c* here are the distinctive feature or the phonological relation of difference in person. Every object is double without it being the case that the two halves resemble one another, one being a virtual image and the other the actual image. They are unequal odd halves" (270–71/209–10). The "order of reasons" that he lists therefore describe the process of actualization itself, 'first' by means of a differentiation of the virtual content itself, and 'secondly' by means of a differenciation that further differenciates the virtual content as actualized. As an example, we can again appeal to the case of the egg, which further brings out the intermediary notions of individuation and dramatisation: "We think that the difference of intensity, as this is implicated in the egg, expresses first the differential relations or virtual matter to be organized. This intensive field of individuation determines the relations that it expresses to be incarnated in spatio-temporal dynamisms (dramatisation), in species which correspond to these relations (specific differenciation), and in organic parts which correspond to the distinctive points in these relations (organic differenciation)" (323/251). In a manner of speaking, there is a progressive series of different/ciations, leading from the virtual to the actual, though this should not be misunderstood as a teleological understanding of progression. Instead, they are unequal odd halves of a process that bears more and more difference, on various levels, such that different bodies of a greater or lesser degree emerge.

The indications given here of the embryonic egg allow us to better appreciate the otherwise incongruous description of the egg in Deleuze and Guattari's *A Thousand Plateaus*. Instead of in comparison to the world, the egg is now more directly affiliated with the virtual field that comes to be called "the body without organs." "The BwO [Body without Organs] is the egg," Deleuze and Guattari write, and again: "The egg is the BwO" (ATP, 202/164). While the terminology changes, as do the ideas underlying them ever so slightly, the basis of Deleuze's earlier ontology applies just as well within this collaborative work.[27] As various commentators have noted, the concept of the body without organs (hereafter referred to by its acronym,

BwO) is often misinterpreted due to its paradoxically clear terminology. The BwO has nothing to do with a corporeal body lacking its organs, as if gutted and splayed out on the coroner's table. Nor is it "a question of a fragmented, splintered body, or organs without the body (OwB)" (ATP, 203/164). The problem with either of the former interpretations is that both still rely on some notion of a prior unity that has been lost, broken, or otherwise disrupted. The BwO is not a fragmented part of any supposed totality. The very idea of a totality or organized whole is not tolerated by the BwO. Rather, the BwO resists organization as such and cries foul when any attempt is made to reify it. It is for this reason that "[w]e come to the gradual realization that the BwO is not at all the opposite of the organs. The organs are not its enemies. The enemy is the organism. The BwO is opposed not to the organs but to that organization of the organs called the organism" (196/158). Why might this be?

The BwO is nearly synonymous with "the field of immanence," "the plane of consistency" (191/154). It is the virtual. What Deleuze is resistant to is the notion of any hierarchical plan that may be imposed on what we might simply call the "plane of Nature," though we must be clear that this refers to no distinction between the natural and the artificial. As opposed to the idea of an hierarchy, with its linear and arborescent order, it would be more accurate to suggest something along the lines of what De Landa calls "self-organized *meshworks* of diverse elements."[28] The BwO, like the egg examined earlier, is an uninterrupted continuum from out of which various "meshworks" assemble through rhizomatic conjugations. The BwO is articulated from within, it has no form or correct structure, it is nonlinear. The organism is considered the enemy precisely because it is traditionally taken for a static (vs. genetic) and organized (vs. rhizomatic) "judgment of God." Against this notion of an order imposed from above, the BwO does not exude pure chaos so much as it is the continuum of differential multiplicities through which intensities are actualized, or, more in keeping with the terminology of A *Thousand Plateaus*, through which they are stratified. It just so happens that the organism proves to be a particularly effective illustration of such a stratification from the BwO egg: "we treat the BwO as the full egg before the extension of the organism and the organization of the organs, before the formation of the strata" (ATP, 190/153).

By dismantling the organism—which Deleuze and Guattari are clear to stipulate is not a haphazard and wild suicidal mission but rather a more careful experimentation—one passes to the plane of immanence, the body without organs. It is on this plane that one is always under way, already farther on, ever moving. It is the site for continual transformation, genetic metamorphoses, and becomings, as opposed to the stratification and organization of static being. A stratum, such as the organism, lodges and roots itself in the BwO. With this notion of the BwO, however, it is important to keep a few things in mind. Not only does its name suggest an "anorganismic" connotation—indicating both a corporeal yet incorporeal sense, a body of a different sort, as seen earlier—it is nevertheless almost always invested with some sort of biological connotation. As already mentioned, there is the egg. The embryological egg is akin to the virtual insofar as it is formless, unordered, transformative, and always in a state of metamorphosis. The virtual cannot be thought as comparable with the "possible" or "potential" (cf. DR, 269–76/208–14). The virtual egg is very real, even if it is not actual. The virtual is almost akin to a quantum position of multiplicities that is actualized through a process of differenciation—that is, through a process of integration. In this respect, we can say that the organism, as our example, is an actual solution to a virtual problem. It is resolved through a genetic process of greater and greater different/ciation.

In addition to the embryonic egg and BwO, there is also "the famous prebiotic soup" (ATP, 66/49) or "prehuman soup" (83/64), that speculative mixture out of which life first emerged. While Deleuze's BwO does not solely pertain to the living—his usage of "machinic assemblages" is precisely a means of working away from a conceptual dichotomy between the living and nonliving, natural and artificial—I suggest that one finds in this notion of "prebiotic soup" yet another reminder of how much his thought is deeply immersed in the field of biology. On this notion of a primordial prebiotic soup, Daniel Dennett reminds us of a letter that Darwin wrote, in which he surmised that he really didn't have a good theory for how life first began, but that perhaps it did so in something like "a warm little pond" (149).[29] Deleuze's own reference is with respect to the "substrata that furnish the materials (a prebiotic soup, a prechemical soup . . .)" (ATP, 73/55) for stratification. These descriptions offer an especially evocative picture of the emergence of distinct beings. Just as biologists and chemists have attempted to determine the first appearance of life from out of the Earth's earliest chemicals such as carbon, nitrogen, hydrogen, and oxygen, as well as various minerals, so too does Deleuze draw a parallel with this prebiotic soup as a kind of substratum from out of which strata emerge. "What organism is not made of elements and cases of repetition, of contemplated and contracted water, nitrogen, carbon, chlorides and sulphates, thereby intertwining all the habits of which it is composed?" (DR, 102/75).

The soupy mixture, however, is not a perfect model for the BwO, nor is it intended as such. On the one hand, Deleuze and Guattari do not appeal to the BwO as a prior stage or level that temporally or spatially precedes strata-like organisms. Whereas the prebiotic soup is often considered a homogeneous and far-from-equilibrium plane that eventually settles into a self-organizing system in which some formal order emerges, Deleuze and Guattari do not consider the BwO to be a part of a developmentalist paradigm, as though organisms represented a progression and state removed from the BwO. As they note, "we should be on our guard against any kind of ridiculous cosmic evolutionism" (ATP, 65/49). Rather, the BwO and the organism coexist as simultaneously real, as though the prebiotic soup is still a very real virtual continuum that gives birth to new actualized life. The BwO, in other words, is a plane immanent to the stratification that occurs on it. This is why life may be considered a complex relation as opposed to a development of form, and a body can be anything as long as it is considered in terms of its set of relations and affects. The BwO is a virtual state—an ontological dimension that is just as real as the actual strata—of intensive and extensive processes, both a plane on which strata-like organisms emerge as well as the ever-changing plane that allows for new becomings, transformations, passages, and assemblages within and between the strata themselves. "The organism is not at all the body, the BwO; rather, it is a stratum on the BwO, in other words, a phenomenon of accumulation, coagulation, and sedimentation" (197/159).

A further departure from the homogeneity of the famous soup is that the BwO is nothing but heterogeneous differences, not only due to the changing face of its plane but also because of the ontological dimension that Deleuze highlights in the univocity of being. If we were to consider the BwO as univocal, for instance, it is so only because it speaks of difference as such. "This body without organs is permeated by unformed, unstable matters, by flows in all directions, by free intensities or nomadic singularities, by mad or transitory particles" (53–54/40). In this manner, the BwO both parallels and departs from the idea of a prebiotic soup, which is perhaps why the egg may still offer the 'best' model of the BwO, even if it does so only allusively.

This discussion of the BwO should begin to highlight what Deleuze is up to with his ontology of the organism. His focus rests less on the static condition of actualized things than it does on how things become actualized. Everything is in a process of becoming; however, this process occurs not only at a level between already actualized things, but rather within the actualizing and differentiating agencies themselves. The organism is not a self-identical substance that can be classified according to formal patterns. Instead, we are asked to recognize the 'molecular' processes implicated within the molar entity that we are more accustomed to dealing with in a more traditionally empirical way. Deleuze's peculiar version of ontology—it has also been called a "transcendental empiricism"—implores one to consider the genetic factors that compose a body, that speak to how a body is made. One can think of these factors as molecular: "In so far as it testifies to individual actions between directed molecules, an organism such as a mammal may be assimilated to a microscopic being" (DR, 329/256). But the organism is not only molecular. As we have still to see, the organism is both molar and molecular, both extensive and intensive, both stratified and on a BwO. These perspectives really only pertain to the ontological level that one addresses, since any given body can and should be observed in terms of the connections, thresholds, and affects that it may undergo, no matter whether these are considered biologically, socially, molecularly, politically, economically, cosmically, psychoanalytically, or so on. There is, as Brian Massumi notes, "the absence of a clear line of demarcation" between descriptive registers.[30] In a way it all comes down to the heterogeneity of life, across which different beings emerge, giving rise to different ways of speaking about them, though all in one and the same voice. As Deleuze and Guattari aptly put it, "What we are talking about is not the unity of substance but the infinity of the modifications that are part of one another on this unique plane of life" (ATP, 311/254).

Hence the need to experimentally dismantle the organism in order to clarify ontologically what makes up *this* particular composition. The account of organisms does not rest with the BwO, however. To be sure, the intensive process involved in the genesis of life is key to Deleuze's ontology. But after having dismantled the organism, so to speak, it might be best to also consider how this being enters into assemblages with other entities, such as with those other things within an environment. For the ontological process does not start and end with the molecularization of beings, particularly since beings such as humans, animals, and rocks are just as much a part of the molecularization of life as the transitory particles of the BwO itself. In other words, in becoming-molecular, we do not simply stop at stratified entities, but continue to observe this ontology at work throughout all of the cosmos.

NATURE'S REFRAIN SUNG ACROSS MILIEUS AND TERRITORIES

We will not reach the crescendo of the cosmos, however. This will have to wait for another time, even if it is necessarily implicit in much of what follows. Instead, I am more interested in remaining with the animals and how they connect with their environments. Deleuze even indicates this direction, and in doing so foreshadows what would be taken up much more explicitly in A Thousand Plateaus: "A living being is not only defined genetically, by the dynamisms which determine its internal milieu, but also ecologically, by the external movements which preside over its determination within an extensity" (DR, 280/216). We could even continue this one step further and suggest that environments, packs, species, and so on, are just as much individual entities as organisms are, it is just that they operate at different scales.³¹ It is toward such an environmental—indeed, even cosmic—view that I wish to move. As we observed earlier, beings such as organisms need to be opened up and dismantled in order to inquire into their 'internal' compositions, into the genetic processes that move from a virtual embryonic state to its actual state of affairs. Phenomena such as organisms are not static organized unities, but porous bodies that assemble into compositions through a variety of relations. These relations can be considered ontologically as intensive forces, as a series of differentiations that accumulate genetically. These relations can also be considered, however, as relations between more extensive entities if considered ecologically. While the perspective changes, the ontological dimensions are just as much in place.

For example, while I considered the BwO as approximating the prebiotic soup from out of which life emerges as a self-ordering process, the same may also be said with respect to the creation of environments or societies. They are governed just as much by the ontological forces of the BwO as organisms are. With this in mind, we discover that "[f]rom chaos, *Milieus* and *Rhythms* are born" (ATP, 384/313). The idea is the same: from out of a chaotic flux, various patterns may emerge. This is one of the main principles behind complexity theory—namely, that complex systems have a tendency to fall into fixed patterns, or that they may, in other words, spontaneously generate order. The order is nevertheless part of a constant and overall movement, so that whatever order arises will be fleeting, both temporally and spatially. This notion similarly relies on the second law of thermodynamics that "states that everything that happens in the universe is accompanied by an increase in entropy," leading toward something akin to a homogenizing chaos.³² Thus, while we might think of order being generated within localized areas (e.g., organisms, weather patterns, ecosystems, stock markets, cyclical time), the overall conversion of energy is leading toward disorganization within the universe as a whole. With respect to the genetic creation of milieus, rhythms, and territories, Deleuze and Guattari want to emphasize that they too are born from 'chaos,' implying that some resemblance of order—albeit an order that we have yet to evaluate—emerges to create these diverse assemblages along a plane of consistency. It is in this manner that we see how chaos is the "milieu of all milieus" (ATP, 385/313), such that chaos finds a rhythm in milieus, where milieus are the becoming rhythmic of chaos, the interlocking of different milieus together *as* rhythm. Against the backdrop of this "chaosmos," and yet immanent to it, diverse regions and pockets of fragile order pass one into the other, giving the appearance of some stability.³³ But the stability is only an illusion.

One such pocket is that of the *milieu*, and it is in this manner that Uexküll returns within our discussion. As we have noted in earlier chapters,

Uexküll emphasizes that all animals live within a circumscribed environment that is peculiar to each animal alone. Their individual environment—that is, their *Umwelt*—is full of significance in accordance with only those signs that register for them. Thus, Uexküll likens the *Umwelt* to a soap bubble encircling each animal, within which certain things are significant and outside of which things simply are not manifest. This is their own subjective domain through which the *Umwelt* reciprocally defines the animal as a subject. With Deleuze and Guattari, however, this notion of a self-enclosed milieu is disrupted and fragmented, to the extent that they claim "the notion of the milieu is not unitary" (ATP, 384/313). Instead of a single bubble encompassing an animal life, milieus now pass one into the other, traverse one another, and can even be divided up into four corresponding milieus for each living being: there is an exterior milieu (e.g., material 'outside' the organism), an interior milieu (e.g., the organs), an intermediary milieu (e.g., the membrane or skin of the porous body), and an annexed or associated milieu (e.g., energy resources such as food, light, and air) (384/313; 65–69/49–52). When characterized in this way, living beings do not simply 'have' these milieus as a location in which they live, but *are* a composition of these milieus. There is a stratification of the milieus in organisms. The flip side of this is that insofar as organisms are a composition of milieus, and milieus slide one into the other, the living being, just like the milieu, does not present a unified whole either. Through this look at milieus, I want to underscore a significant difference between Deleuze and Guattari's interpretation of milieus—which nearly always feature Uexküll—and the readings by Heidegger and Merleau-Ponty. Of greatest interest, I believe, is how the unity of the milieu is "shattered" into diverse parts. Needless to say, this bears on the stratified organism as well. In other words, neither the milieu nor the organism is as coherent as we would like to believe. Consider the following, for example: "A stratum, considered from the standpoint of its unity of composition, therefore exists only in its substantial epistrata, which shatter [*brisent*] its continuity, fragment its ring [*l'anneau*], and break it down into gradations. The central ring does not exist independently of a periphery that forms a new center, reacts back upon the first center, and in turn gives forth discontinuous epistrata" (67/50–51).[34] With Deleuze and Guattari, therefore, the soap bubble has burst, with various repercussions.

The ring just mentioned is not explicitly discussed in connection with Uexküll's soap bubble per se, and yet their discussion of both Uexküll and milieus is situated between these passages on the shattering of rings. It is remarkable that Deleuze and Guattari's choice of words should match so well with that of Heidegger's encircling rings and Merleau-Ponty's rings of finality, both of which coincide with their respective readings of Uexküll. Moreover, even though Heidegger will speak of penetrating an animal's encircling rings, and Merleau-Ponty will refer to the overlapping and crossing of concentric circles, neither gives full credence to the shattering of the animal's ring. We recall that Heidegger will occasionally refer to a shattering and rupture, as well as the intermeshing of the encircling rings, but even these do not carry the same degree of instability—or, at any rate, the same repercussions—that Deleuze and Guattari suggest by their usage. The shattering of the strata sounds violent, but it isn't any less natural. At stake is the natural movement that associates a stratum (e.g., an animal) with its milieu, but in such a manner that both stratum and milieu are decentered one with respect to the other yet nevertheless a part of a certain rhythm.

A milieu, as characterized by its fourfold structuration, is "a block of space-time" that is "coded" through periodic repetition in such a way that this block becomes significant for a certain animal. But in saying this, it is just as much the case that milieus can be decoded and recoded—a transcoding—in accordance with the living being in question. Deleuze and Guattari cite genetic mutation and viruses as examples of this phenomenon, whereby a code in one milieu (e.g., a gene or a virus in one population of species) may be transcoded by a different species (69–70/53). Research in biogenetics operate on precisely these lines, whether in terms of the genetic engineering of vegetables or the vaccinations made available for humans through genetic research done on mice. Fragments of code are read differently according to the being in question. The point here is that Deleuze and Guattari see the milieu as a coded material domain for an organism receptive to repeated elements. But we must not forget that the concept of milieu is not only applicable to living beings; there are milieus at greater and lesser dimensions. In the case of transcoding, there is the constitution of a new linkage between living beings. This is how milieus are said to overlap and pass from one into another, insofar as their elements may be transcoded, and, in effect, to communicate with one another. Such communication proceeds due to the rhythm that is located between different milieus. So, for example, Ronald Bogue explains how the breathing rate of a mammal might be in rhythm with the beating of its heart, its hormonal stimuli, its ability to transpire, and alterations in the environment.[35] Each milieu—interior, exterior, intermediary, annexed—communicates with another as a rhythm that passes through them.

As witnessed by the chapter heading "Of the Refrain" in *A Thousand Plateaus*, this discussion of milieus, as implicated as they are with rhythms, songs, and vibrations, is situated in terms of music. It should come as no surprise, then, that in correspondence with our movement back toward animal ecology and ethology, Uexküll is once again a focus of attention. And yet, there is a significant omission: it should not be overlooked that in Deleuze and Guattari's discussion of milieus—wherein Uexküll features decisively—the

Umwelt is never raised as such. Whereas the *Umwelt* featured in association with both Heidegger's and Merleau-Ponty's interpretations, it is never thematized by Deleuze and Guattari as such. This was not an oversight, to be sure. Had they wanted to appeal to the *Umwelt*, they most certainly would have. An indication of this is in the sole mention of the *Umwelt* in *A Thousand Plateaus*, and it is mentioned in relation to a distinction between animal *Umwelten* and a scientific *Welt*, but it is raised neither in connection with Uexküll nor as a conceptual aspect of their thought (81/62). Either way, it is not thematized as such. It is as if they purposely eschew a discussion of the most obvious aspect of Uexküll's thought: his theory of the *Umwelt*. What we have instead is a discussion of the affective body (e.g., the tick's three affects), as well as, we now discover, a specifically musical connotation of nature. There is a rhythm that passes between milieus as well as a greater melodic landscape that connects different stratified bodies.

The composition of Nature, which was tied earlier with the plane of immanence, is now raised in connection with music. Within this context, Uexküll is praised for elaborating a theory of "Nature as music" from which we observe various transcodings that operate between organisms. For example, "It has often been noted that the spider web implies that there are sequences of the fly's own code in the spider's code; it is as though the spider had a fly in its head, a fly 'motif,' a fly 'refrain.' The implication may be reciprocal, as with the wasp and the orchid, or the snapdragon and the bumblebee" (386/314). The reciprocity highlighted here is that which we have already considered in our earlier treatments of Uexküll. The melodic composition, or the harmony between different living beings, is contrapuntal, such that the spider connects with the fly in composing not only a new relation but an altogether "new plane." This is when Deleuze and Guattari's reading acquires new force. In comparison to Heidegger and Merleau-Ponty, Deleuze and Guattari show little interest in the environmental world of animals. This may seem questionable in light of their discussions of nature, milieus, and, as we have still to see, territories. Yet these conceptualizations have less to do with an environmental space-time within which an organism lives than with how beings enter into various relations with other things. It is the constitution of new planes, in other words, rather than in what an organism is or is not open to, that is ontologically decisive for them. In their final collaboration, *What Is Philosophy?*, Deleuze and Guattari note that "these relationships of counterpoint join planes together, form compounds of sensations and blocs, and determine becomings" (176/185), and it is precisely the nature of these becomings that is at stake throughout this ontology.

But before going further into the concept of becoming, wherein we find the great example of the orchid and the wasp and its connection to Uexküll, we need to observe milieus in relation to territories and refrains. So far, we have been considering milieus and the rhythms that pass between them. These milieus, however, are said to be directional and functional because they involve, in one way or another, material: "materials, organic products, skin or membrane states, energy sources, action-perception condensates" (ATP, 386–87/315). Furthermore, these materials are coded in such a way that they register with other codes (a spider code with a fly code) and likewise may be transcoded such that they function in different manners according to different beings (a tick code receives a particular mammalian code, but not a fly code). But despite these associations, milieus are still fundamentally materials that become encoded through periodic repetition. *Territories*, in contrast, are created through an expressive act that affects milieus. Whereas milieus may be considered materialistic (though by no means mechanistic), territories are established by the marking of a "signature" that territorializes a milieu. A territory is the product of territorialization, which itself is "an act of rhythm that has become expressive, or of milieu components that have become qualitative" (ATP, 388/315). Ronald Bogue helpfully interprets this territorialization as the "autonomy of qualities and rhythms" (20), which is also a way to discern that every territory is an unconstrained act of deterritorialization and reterritorialization. Every deterritorialization is accompanied by a subsequent reterritorialization, a marking of a milieu that is expressive and not solely subservient to a natural direction or function. For example, a monkey might expose its brightly colored sexual organ to mark a territory that it is guarding, a rabbit might urinate around a field to communicate a perimeter, and a bird might drop leaves from a tree in order to express its territory (ATP, 387/315). Similarly, a human might plant a flag or paint a cave to territorialize a particular domain. In each case, the living being marks a signature with an expressive act that extends beyond any direct correlation to a specific type of action; there is a note of autonomy to the act. These examples therefore ought to be considered differently from an instinctual display, such as how a fish might change color to show aggression. A sign such as this remains functional and action-specific, and thus tied to a milieu. A territory, on the other hand, is extracted from a milieu because its expressivity breaks away from the coding specific to its milieu. In these examples, the sexual organ, urine, leaves, and paint de- and reterritorialize an otherwise stratified milieu through acts of expression.

A further mode of distinction is that while there is a stratification of milieus (e.g., the organism), there is a comparable manner of thinking of territorialized assemblages (e.g., wasp-orchid). The marking of territories is inseparable from both a deterritorializing process and subsequent reterritorializations, both of which are enacted through an assemblage of intensive processes. Deterritorialization is akin to a "line of flight," an escape from one domain to another. For example, the monkey's sexual organ is deterritorialized as a sexual organ and reterritorialized in becoming a display of guardianship. Or, better yet, we could speak of the de- and reterritorialization that occurs between one of Deleuze and Guattari's recurrent examples, that of the orchid and the wasp:

> The line or block of becoming that unites the wasp and the orchid produces a shared deterritorialization: of the wasp, in that it becomes a liberated piece of the orchid's reproductive system, but also of the orchid, in that it becomes the object of an orgasm in the wasp, also liberated from its own reproduction. A coexistence of two assymetrical movements that combine to form a block, down a line of flight that sweeps away selective pressures. The line, or the block, does not link the wasp to the orchid, any more than it conjugates or mixes them: it passes between them, carrying them away in a shared proximity in which the discernibility of points disappears. (360/293–94)

The reciprocal line of flight that passes between the two terms—and we should note that the emphasis is not on the wasp and orchid as such, but on the process between them—simultaneously unites them while at the same time, as a passage, carries them both off in different directions. Together they are implicated in an "aparallel evolution" (swept away from selective pressures) because there is no possible filiation between the orchid and wasp. They do not reproduce with one another, but instead de- and reterritorialize the other. Similarly, this connection could be better characterized as "symbiotic" since it brings together different organisms that coexist in a singular becoming, "but from which no wasp-orchid can ever descend" (291/238). When understood as aparallel, symbiotic, a block of becoming that produces a line of flight that traverses territories and milieus, the orchid and the wasp can be said to form a "rhizome" (17/10). The rhizomatic assemblage that this pair forms is more like a hybrid than any specific or generic entity.

I would also add that the orchid and the wasp engage in a becoming that is specific to this passage between them. In this process of de- and reterritorialization between the two, "something else entirely is going on: not imitation at all but a capture of code, surplus value of code, an increase in valence, a veritable becoming, a becoming-wasp of the orchid and a becoming-orchid of the wasp. Each of these becomings brings about the deterritorialization of one term and the reterritorializaiton of the other; the two becomings interlink and form relays in a circulation of intensities pushing the deterritorialization ever further" (17/10). I want to emphasize this illustration of the orchid and the wasp because of just how similar it is to Uexküll's own example of the flower and the bee that he raises in *The*

Theory of Meaning, a text with which Deleuze was familiar. In this text, Uexküll develops the example of how a bee and a flower approximate one another in a meaningful confluence of signification (TM, 65). Each finds a complement in the other: the flower is described as bee-like and the bee as flower-like. Together they form a harmonious duet that meshes the two together. While Uexküll does not advance an ontological position for himself, I suggested before, and wish to reiterate again, that this coupling proves to be a significant model from which an ontology may emerge. So far as I know, Deleuze and Guattari's example of the orchid and the wasp is not attributed to any particular source, despite the copious footnoting found in *A Thousand Plateaus*. And while it matters little whether Deleuze and Guattari's example of the orchid and wasp are a repetition of Uexküll's flower and bee, the parallel between the two nevertheless underscores a significant dimension to the ontology that Deleuze and Guattari elaborate.[36] The orchid and the wasp form a rhizomatic becoming that exists as an assemblage between heterogeneous terms. Uexküll's own example does not evoke anything near the dynamism that Deleuze and Guattari read into the relation; with Uexküll, the orchid and wasp would still be caught within their own bubbles, albeit in a manner in which each is significant for the other. With Deleuze and Guattari, on the other hand, the bubbles have burst due to the lines of flight that carry each term off in new directions. The orchid and wasp become nearly synonymous with the ontological break effected by this line of thinking.[37]

"But," Deleuze and Guattari note, "it is not just these determinate *melodic compounds*, however generalized, that constitute nature; another aspect, an infinite *symphonic plane of composition*, is also required: from House to universe" (WP, 176/185). Just as there is a rhythm that passes between milieus, there is a *refrain* that belongs to territorial assemblages. The example that is often noted here is one that Deleuze and Guattari pull from studies in ethology: a bird's song. The song of a bird marks its territory and does so in three distinctive ways, each relating to different aspects of the thought we have been describing thus far. The three aspects of the refrain include, first: "The song is like a rough sketch of a calming and stabilizing, calm and stable, center in the heart of chaos.... the song itself is already a skip: it jumps from chaos to the beginnings of order in chaos and is in danger of breaking apart at any moment" (ATP, 382/311). This corresponds to a territorial assemblage. The song's rhythm is born from chaos like a cry from the dark. Second: "Now we are at home. But home does not preexist: it was necessary to draw a circle around that uncertain and fragile center." This corresponds to an intra-assemblage that has constructed some order by the fashioning of a home. Third: "Finally, one opens the circle a crack, opens it all the way, lets someone in, calls someone, or else goes out oneself, launches forth. One opens the circle not on the side where the old forces of chaos press against it but in another region, one created by the circle itself." This corresponds with an interassemblage, a passage, or line of flight, leading beyond the organized home. Bogue, who has given these passages some attention, summarizes the three aspects as "a point of stability, a circle of property, and an opening to the outside" (17). Of particular interest is not only how the musical motif of the refrain is employed, but how it is done in keeping with both Deleuze's approximation of complexity theory, as well as in how this language of being at home by drawing a circle around oneself evokes the earlier accounts of animal environments seen in Heidegger's "encircling rings" and Merleau-Ponty's "rings of finality." The manner by which the refrain is described is particularly evocative, especially after the language we have already observed in both Heidegger and Merleau-Ponty. With Deleuze and Guattari, however, we discover a more deliberate cracking of the circle and less stability in its structure. As we already observed in the case of the fractional milieus, the refrain also demonstrates the necessity to pass beyond the calming circle by cracking it open in a line of flight. Out of the midst of chaos, a single voice or song arises, takes root, territorializes a home by circling itself, only to pass on in a new movement of deterritorialization. Every territory is not without its de/reterritorialization, just as every milieu is not without its de/recoding, and every stratum is not without its de/restratification.

And yet, where is the refrain in all of this? What are we to make of this musical composition that unfolds through all of Nature, from rhythmic milieus and melodic counterpoints to the refrain passing through it all? Bogue asks precisely this question, and does so in an intriguing fashion: "Where is the refrain? . . . Is the refrain in the tick or in the mammal whose blood it sucks, in the spider and its web or in the fly for whom the web seems so specifically designed? The refrain is the differential rhythm constituted in milieus, the relation between milieu components, and though one can speak of the melody of the octopus and the countermelody of the water, their contrapuntal *relation* is the refrain, one that belongs to both but in a sense to neither" (74). This formulation of the problem is especially intriguing in that it mirrors a description we have already seen: isn't this exactly how Merleau-Ponty formulates his interpretation of Uexküll? Merleau-Ponty determines Uexküll's melodic *Umwelt* to be that which sings through us, immersing the body in its thickness, such that neither the animal nor its milieu is the generative cause of either, but suspended together in this contrapuntal relation. This turned out to be an initial approximation of Merleau-Ponty's theory of the chiasmatic flesh—an ontological concept that addresses the reversibility between bodies in the envelopment of life. But surely this is not what Deleuze and Guattari mean, is it? There must be something different to their interpretation. If this is the case, however, then what do we gain from these insights?

Perhaps more than anything else, it is the idea that the refrain exemplifies the relations between milieu, territories, and the deterritorializing lines of flight that allow us to focus on the passages as such. For isn't this what the cracking of the circle, the shattering of strata, the opening onto other planes is all about? Everything, whether an egg, an organism, a milieu, a territory, an earth, or the cosmos, is at once a calming and stabilizing of forces while at the same time a passage elsewhere, a relation between, a becoming of something different. It is not the organism that is of interest since there is no organism *as such*. There is only ever a genetic becoming, a processional heterology. The distinctions between 'internal' and 'external' apply only as long as we continue to think of the organism as the point around which everything else is organized. But, as we have observed, this point is fragile due to the nature of its composition. Instead of the biology of the organism, we are instead asked to consider an ethology of affects.[38] With the refrain, there is a passing, an assemblage that is always a becoming, "the song of the virtual, " as Badiou puts it (48). It is with this concept of becoming that I wish to conclude.

THE ANIMAL STALKS

Aside from the previous considerations, it is still the status of the animal organism that I am interested in. Yes, it has been named the enemy. And, granted, it poses an unsatisfactory solution to complex problems. And yes, it becomes de- and restratified, de- and recoded, de- and reterritorialized, fractured and shattered, and that it does so at various ontological levels, whether molecularly or macroscopically, affectively or territorially, and so on. Yet it is the animal that is nevertheless the focus here, right to the end. In connection with the concept of becoming, it would seem that we must at least take a brief pause with the concept of becoming-animal. While this has little do with a re-instantiation of the concept 'animal' or 'organism'—as though the becoming-animal was some sort of resurrection of the animal out of its shattered images—we still have to contend with the actuality of living beings. It is with the question of onto-ethology that I wish to conclude.

Roughly midway through *A Thousand Plateaus*, Deleuze and Guattari write the following: "Climate, wind, season, hour are not of another nature than the things, animals, or people that populate them, follow them, sleep and awaken within them. This should be read without a pause: the animal-stalks-at-five-o'clock" (321/263). This is perhaps an improbable place to end, but in many respects it cuts to the chase in our consideration of Deleuze's ontology. As expressed in the forthright suggestion that one must read "the animal-stalks-at-five-o'clock" without pause, Deleuze and Guattari highlight one of the principal components underlying this ontology, namely, that of an individuality immanently composed through a connection of diverse affects. Of particular note, it is worth observing how similarly, yet how differently, this phrase repeats one of the most recognizable expressions in contemporary ontology. I am thinking here of Heidegger's introduction of his neologism "being-in-the-world," where he is just as clear as Deleuze and Guattari are with respect to the necessity of regarding this notion as a whole. Heidegger writes: "The compound expression 'Being-in-the-world' indicates in the very way we have coined it, that it stands for a *unitary* phenomenon. This primary datum must be seen as a whole. But while Being-in-the-world cannot be broken up into contents which may be pieced together, this does not prevent it from having several constitutive items in its structure" (GA2, 53/78). Just as Heidegger leans on the hyphens in being-in-the-world to indicate this unitary phenomenon of being, Deleuze and Guattari suggest just as much with their own unique phenomenon. The animal-stalks-at-five-o'clock has an inherent, one should even say intensive, immediacy to it, such that the contents cannot be broken up into parts unless one wanted to risk changing the very composition in question. To pause in this expression, you change the nature of the consistency that holds this individuality together: "to break the becoming-animal all that is needed is to extract a segment from it" (ATP, 318/260). Held in balance, then, is that "Five o'clock is this animal! This animal is this place!" (321/263). We have already observed how Heidegger's ontological expression plays out in the midst of animal environments—namely, how being and world are accorded, or, in the case of world, not accorded, to animal life. As a final word, it will be Deleuze's ontology that becomes an issue, specifically in such a way that we may better understand how this simple phrase—the animal-stalks-at-five-o'clock—captures the thought of becoming-animal that is so central to his thought.

We would do well to appeal here to the concept of "haecceity." Earlier we saw that Deleuze invoked Duns Scotus's univocity of being in order to speak to the difference within all of being; he appeals to Scotus again with the term haecceity. The concept of haecceity derives from Scotus's reflections on the individuality or singularity of "this thing [*haec*]." In contrast to the 'whatness' or quiddity of something, Scotus raises the need to address the 'thisness' or haecceity of an entity's singularity. Haecceity is therefore raised as an alternative to the essence or generality of a thing by instead focusing on the individuality of a given body. Another way of differentiating the two is to state that quiddity refers to substance, whereas haecceity refers to the modes of individuation or affects that are distinct from substance or subject. Deleuze and Guattari write: "There is a mode of individuation very different from that of a person, subject, thing, or substance. We reserve the name *haecceity* for it" (318/261). To speak of a climate, a wind, an hour, or a place requires that one address this perfect individuality that composes each assemblage. The climate and hour "are haecceities in the sense that they consist entirely of relations of movement and rest between molecules or particles, capacities to affect and be affected" (318/261; cf. 310/253). What is noteworthy in Deleuze and Guattari's employment of the term "haecceity" is that it is used to describe the intensive processes that compose a given body, all the while being irreducible to *this* body, as Keith Ansell Pearson has shown (181). Haecceity, in other words, refers to the singularizing process that belongs to every body in its state of becoming, inasmuch as a body is always in movement in its capacity to affect and be affected. Haecceity points to the singular *thisness* of any and every composition.

What does this mean? For one, that "you will yield nothing to haecceities unless you realize that that is what you are, and that you are nothing but that" (ATP, 320/262). A being is not a subject, on the one hand, and a series of modes or accidents, on the other. Nor are haecceities the backdrop or Gestalt against which subjects emerge. A body is, as we have seen, its ability to affect and be affected—that is, its process of becoming. This is in part why Deleuze and Guattari note that the weather, hour, or place are inseparable in nature from things, animals, or people. They are of the same nature, and being must be said univocally. The animal-stalks-at-five-o'clock, therefore, is the singularity of *this* haecceity, of this assemblage. Inasmuch as one looks to these affective becomings, Deleuze's ontology can be characterized as an "onto-ethology," albeit in this specialized sense. One needs to count affects and map a body like that of an animal-becoming-night or a cloud-of-locusts-swept-by-the-wind.

> It is the entire assemblage in its individuated aggregate that is a haecceity; it is this assemblage that is defined by a longitude and a latitude, by speeds and affects, independently of forms and subjects, which belong to another plane. It is the wolf itself, and the horse, and the child, that cease to be subjects to become events, in assemblages that are inseparable from an hour, a season, an atmosphere, an air, a life. (321/262)

The animal-stalks-at-five-o'clock is just a particularly forceful manner of depicting this point. It is not a matter of describing the organization of an animal. Nor is it a question of following its line of descent, whether developmentally (as ontogeny) or evolutionarily (as phylogeny). Nor is it a case of imitating animal life in an attempt to become like them. Rather, it is a question of following the disparate becomings of any and all individuated things. Deleuze's thought decenters the animal—and not just the animal, but all stratified, extended entities—as the focal point for investigation. The animal is not the center around which the assemblage "the animal-stalks-at-five-o'clock" turns. The hour, the weather, hunger—it all goes into this assemblage. A becoming, therefore, is the creation of a new assemblage where affects collate to produce a new body by passing from an actual state of affairs through a virtual field toward a new actualization.[39] This pertains to all of life: "If everything is alive, it is not because everything is organic or organized but, on the contrary, because the organism is a diversion of life. In short, the life in question is inorganic, germinal, and intensive, a powerful life without organs, a Body that is all the more alive for having no organs, everything that passes *between* organisms" (623/499). This is true not only for living beings, nor even for all of life, but across and between the univocity of being.

The animal-stalks-at-five-o'clock is about as precise as one can get in revealing the nature of an animal. Pure animality is liberated from the constraints of the organic to become inorganic or supraorganic. Every organism is inseparable from the relations or affects of which it is capable, and to extract the organism from out of its dynamic state would entail losing the very individuality that defines it. A peculiarity of this approach, however, is that Deleuze never really thematizes the concept of 'world,' even after he notes that "an animal, a thing, is never separable from its relations with the world" (SPP, 168/125). Unlike Heidegger and Merleau-Ponty, the concept of world is never entertained as an important part of Deleuze's ontology. To be sure, the world is raised now and again, but just as we observed in his treatment of Uexküll's *Umwelt* (namely, the lack of any discussion of this concept), the world is not at issue either. It is not a relation with the world *as such*, therefore, but a relation with everything that might be found within a virtual plane capable of becoming actualized. Thus, it is the passing that occurs between virtual intensities and their actualizations into individuated beings that is of greater importance. By approaching beings in this way, Deleuze avoids the need to address the openness of an environment or a world to the animal, even though we do see that he emphasizes, just as Heidegger and Merleau-Ponty before him, the process and movement of life. It is just that the ontological relations are different. This is why I believe Uexküll proves to be a useful proponent within Deleuze's onto-ethology. Through Deleuze's readings, we are led to look away from animals themselves toward what they are capable of affectively. Whether it is in terms of a tick's affects, or the contrapuntal relations of a refrain relating an orchid and a wasp, or some other such assemblage, our focus is trained toward what escapes the nodal terms in order to map their becomings. This is the task of a new ethology. " 'Ethology' then can be understood as a very privileged molar domain for demonstrating how the most varied components (biochemical, behavioral, perceptive, hereditary, acquired, improvised, social, etc.) can crystallize in assemblages that respect neither the distinction between orders nor the hierarchy of forms" (ATP, 414–15/336). In Deleuze's onto-ethology, and in his readings of Uexküll more specifically, it is a matter of counting affects because even if there were a concept of world, it would be just as subject to fragmentation as the crystallized assemblages that may compose it. Rather than looking at an animal's behavior, which would presuppose some ordering within an environmental space, we are instead asked to map the affects that make *this* body on the plane of nature. In becoming a good ethologist, Deleuze transforms philosophy and shows how he himself becomes-tick: "The philosopher is no longer the being of the caves, nor Plato's soul or bird, but rather the animal which is on a level with the surface—a tick or louse" (LS, 158/133). But this is just a transitional state too. It is the animal-stalks-at-five-o'clock rather than a being-in-the-world of Dasein. This approach coincides with a reenvisioning of ethology, and it just so happens that Uexküll is named one of its founders.

CONCLUSION

Uexküll and Us

The concluding chapter to Deleuze's little book on Spinoza is playfully entitled "Spinoza and Us." While he explains that this phrase could mean many things, he seems drawn to the idea of us being within the "milieu" or "middle" (*milieu*) of Spinoza. By this he suggests that, in order to understand a thinker like Spinoza, we need to approach him or her not solely by their first principle but also according to their second, third, fourth, and so on. Only thus do we latch on to the plane of their thinking and insert ourselves within their midst. It is my contention that each of the philosophers considered in this book has done precisely this with Uexküll: they have each installed themselves on his plane of thought, they have inserted themselves within his milieus, and that they have done so by approaching his thought in different ways. This is Uexküll and us.

We began at the outset with Uexküll's invitation to explore many different worlds. His stroll through the environments of animals asked us to step out of ourselves, if only for a moment, so as to view our surroundings from perspectives other than our own. His desire was seemingly innocent enough, yet it broke from an anthropocentric attitude that had until then permeated the natural and social sciences. The animal, for one, is not the mechanical object that many would have it be. Likewise, the world is neither a purely objective entity laid bare by natural laws, nor is it necessarily the product of human experience alone. Instead, we are asked to consider the idea that there are as many worlds as there are living beings. Every animal is its own subject, he writes, and so too does every animal create its own subjective universe. One of the implications of this approach is that the reality of the world is meaningful only as subjective experience. This has nothing to do with "animal psychology," however. Uexküll does not investigate mental processes or entertain issues of beliefs, feelings, or consciousness. Had this been his focus, his appeal would have been far more limited. His studies instead point toward behavior in an environment as the means for understanding what it means to be an animal. So by inviting us along for a stroll through the various environments of animal life, Uexküll has in effect revealed a plurality of ethological relations between animal and environment that in turn give expression to something approaching a kind of ontology. The being of animals—how they reveal themselves as intertwined with the environments they in turn create—is expressed through their behavior. To understand what it means to be an animal therefore requires that we understand its relation to an environment. Uexküll leads us in the direction of onto-ethology.

With his interest in the worlds and environments of living beings, it should come as no surprise that Uexküll's thought caught the interest of philosophers. In the preface to *Phenomenology of Perception*, Merleau-Ponty claims: "True philosophy consists in relearning to look at the world" (PhP, xvi/xx). Uexküll not only enacts this very idea, he does so in a fairly unconventional manner. He pushes our natural wonderment at the world one step further. Jump into the soap bubbles of animal life!, he says. Look at how they relate to their surroundings! Every living being reveals a different environment inasmuch as they all relate to other things in an infinite number of ways. It is a matter of these relations, that spells out the being of the animal. Heidegger, Merleau-Ponty, and Deleuze each latch on to this insight and take it in new directions. Part of what makes their respective engagements with Uexküll so interesting is that they each do so over the length of their philosophical careers and that they do so relatively unaware of the other's reflections. Heidegger's engagement with Uexküll spans some forty years, from the 1920s through to his 1967 Heraclitus seminar. Merleau-Ponty demonstrates an interest from *The Structure of Behavior* through his late lectures on *Nature*, and Deleuze likewise references Uexküll's tick from his early writings and lectures through to *What Is Philosophy?* They have each sustained a prolonged relationship with Uexküll's body of thought, and they have done so, moreover, without explicit awareness or recognition of the others' impressions. And yet it is nevertheless around Uexküll that Heidegger, Merleau-Ponty, and Deleuze engage in a silent dialogue. Their ontological positions, though by no means limited to the preceding considerations, suggest a means of comparison in the environment of the animal. They all address ethology in a manner that speaks to their respective projects as a whole. Not solely as a side interest, but in a way that the being of the animal lodges itself in the development of their ontologies.

For his part, Heidegger is drawn to Uexküll because he finds in him an ally in the biological domain. Unlike with other biologists, Uexküll's emphasis on how animals behave in their environments was precisely the kind of ontological equivalent to Dasein's being-in-the-world that would attract

Heidegger. The question that informs his look at living beings is therefore also that of the world: do animals have a world and, if so, how? The question is really whether animals *exist*. Do they reveal an opening onto the world wherein beings can be disclosed in their being? As his primary thesis suggests, animals may indeed have relations within their environments, but these relations do not extend to an understanding of being, resulting in the claim that they are poor in world. Animal behavior demonstrates a manner of being "captivated" by things as opposed to the unbound and open stance of human comportment. The ontological difference between human Dasein and animal life reveals itself in this essential distinction. It is in ethology, therefore, that the being of the animal shows itself. In comparison, it is the transcendent character of human Dasein that allows humans to leap free from the constrictive 'ring' enclosing animal being. In terms of Heidegger's fundamental ontology, his treatment of animal life serves to strengthen our understanding of being-in-the-world while at the same time provides further insight into the problems posed by the body and the environment.

It is also Uexküll's theory of the *Umwelt* that occupies the interest of Merleau-Ponty. But Merleau-Ponty's attention is directed not so much toward the openness of a world as it is toward the body's immersion within its "thickness." What he draws out of his reading of Uexküll is how the environment "unfurls" through animal behavior as a reciprocated active–passive relation. Both the animal and environment are the product of each other's creation in that they are swept together like a melody that suspends the body within the thickness of the world's flesh. Neither the animal nor the *Umwelt* is the focus. Instead, it is the movement of behavior itself that reveals the key to an animal's manner of being. Emphasizing the theme of the melody, Merleau-Ponty moves in the direction of an ontology of nature that dwells on the intertwining between beings rather than on the beings themselves. Behavior is described as sunk into corporeal being to disclose a view of life that is one of process and action, not things and objects. The animal becomes a theme suggestive of movement itself, like a pure wake, unfolding together with its melodic counterparts. The animal is no longer a static substance, but a structure within the leaves of being.

With Deleuze we discover yet another approach to Uexküll, albeit one that offers a fairly different alternative to those of Heidegger and Merleau-Ponty. As opposed to the latter two, Deleuze is not really concerned with the ramifications of the world. It is still too reminiscent of its tie to the phenomenological body. So although he claims than an animal is inseparable from its relations with its world, it is never really the world as such that he investigates. Instead we find a fairly unconventional usage of Uexküll's writings on animal life. Here it is not a matter of being in the world but the becoming of the world. The organism is not the example of life, but its imprisonment. The organism is therefore the enemy in Deleuze's ontology. But like Heidegger and Merleau-Ponty, Deleuze still places an emphasis on the relations between beings; it is just that the way these relations are described differs greatly. It is no longer a matter of describing how the animal behaves in its environment but of counting the affective relations or states of becoming between bodies. Bodies are always emergent properties, some faster and others slower, so that Deleuze's ontology merges with an unlikely source: ethology as the science of affects, of becomings. Thus, rather than thinking of animal lives in terms of strictly defined patterns of embryology or behavior, Deleuze finds in Uexküll a fellow Spinozian ethologist already engaged in counting the affects of animal becomings. Whether this is addressed at the level of molecules, sensations, organisms, milieus, or territories, being can be said univocally across the plane of nature. The rings and circles that previously enclosed the being of the animal in Heidegger and Merleau-Ponty have now shattered and broken in various lines of flight.

Deleuze's emphasis on becoming is not absent in Heidegger or Merleau-Ponty. They have each expressed their reservations with the concept of "organism" and its applicability to ontology. But in its place they have also expressed different versions of animal life. Heidegger suggests that life isn't so much defined by the organism as it is by process and motion. Merleau-Ponty ascribes life to something like a field of action within the hollows of being. Deleuze notes that a life is everywhere as absolute immanence. In the development of their respective ontologies, they have each had the occasion to encounter an unlikely source in Jakob von Uexküll, who simply asked that we think differently about animal life. In response, they each offer a fruitful way of considering the relations between ontology and ethology. It is as though they have gathered within the environments of Uexküll—along his plane of thought—and each proclaimed: this is Uexküll and us.

Notes

INTRODUCTION

1. Uexküll appears in several places in Heidegger, Merleau-Ponty, and Deleuze, as the following study will show. As for the others, references include: Ernst Cassirer, *An Essay on Man: An Introduction to a Philosophy of Human Culture* (New Haven: Yale University Press, 1972); Cassirer, *The Problem of Knowledge: Philosophy, Science, and History Since Hegel*, trans. William H. Woglom and Charles W. Hendel (New Haven: Yale University Press, 1978); Hans-Georg Gadamer, *Truth and Method*, trans. Joel Weinsheimer and Donald G. Marshall (New York: Continuum Press, 1996); José Ortega y Gasset, "Preface," to Jakob von Uexküll, *Ideas para una concepción biológica del mundo*, trans. R. M. Terneiro (Madrid: Calpe, 1934); Jacques Lacan, "The mirror stage as formative of the function of the I as revealed in psychoanalytic experience," *Écrits: A Selection*, trans. Alan Sheridan (New York: W.W. Norton, 1977); Georges Canguilhem, *La connaissance de la vie* (Paris: J. Vrin, 1952); and Giorgio Agamben, *The Coming Community*, trans. Michael Hardt (Minneapolis: University of Minnesota Press, 1993); Agamben, *The Open*, trans. Kevin Attell (Stanford: Stanford University Press, 2004).

2. Marjorie Grene, *Approaches to a Philosophical Biology* (New York: Basic Books, 1965); Hans Jonas, *The Phenomenon of Life: Toward a Philosophical Biology* (New York: Harper & Row, 1966); and Michel Foucault, "Introduction" to Georges Canguilhem, *The Normal and the Pathological*, trans. Carolyn R. Fawcett (New York: Zone Books, 1991).

3. Keith Ansell Pearson, *Germinal Life: The Difference and Repetition of Deleuze* (New York: Routledge Press, 1999); Cary Wolfe, ed., *Zoontologies: The Question of the Animal* (Minneapolis: University of Minnesota Press, 2003); John Llewelyn, *Seeing Through God: A Geophenomenology* (Bloomington: Indiana University Press, 2003); Mark Bonta and John Protevi, *Deleuze and Geophilosophy: A Guide and Glossary* (Edinburgh: Edinburgh University Press, 2004); Charles S. Brown and Ted Toadvine, eds., *Eco-Phenomenology: Back to the Earth Itself* (Albany: State University of New York Press, 2003); H. Peter Steeves, ed., *Animal Others: On Ethics, Ontology, and Animal Life* (Albany: State University of New York Press, 1999); and Babette Babich, Debra Berghoffen, and Simon Glynn, eds., *Continental and Postmodern Perspectives in the Philosophy of Science* (New York: Aveburg Press, 1995).

4. "Onto-ethology" is a term that I borrow from the work of Eric Alliez, whose book *The Signature of the World* offers an interpretation of Deleuze and Guattari's final

collaborative work *What Is Philosophy?* But in pulling my title from one of Alliez's concepts, I must emphasize that my version of onto-ethology has in fact very little to do with Alliez's own usage, which leans exclusively on Deleuze and Guattari, and derives solely from Deleuze and Guattari's one book. Behavior is never really raised by Alliez, and Uexküll not at all. So while I wish to acknowledge that my title derives from Alliez, my own usage is far different in scope.

CHAPTER 1: JAKOB VON UEXKÜLL'S THEORIES OF LIFE

1. A bit more of a mouthful is the translation "species-specific modeling system" by Thomas Sebeok in *Signs: An Introduction to Semiotics* (Toronto: University of Toronto Press, 1994). This translation of *Umwelt* is more of a precise formulation within biosemiotics. I will address biosemiotics below.

2. Kalevi Kull, "Jakob von Uexküll: An introduction," in *Semiotica* 134–1/4 (2001): 9. Biographical references are also drawn from Thure von Uexküll, "Introduction: The sign theory of Jakob von Uexküll," in *Semiotica* 89–4 (1992): 279 and Gudrun von Uexküll, *Jakob von Uexküll: Seine Leben und seine Umwelt, Eine Biographie* (Hamburg: Christian Wegner Verlag, 1964). Thure von Uexküll (1908–2004) was the son of Jakob, and was a university professor of medicine with interests in biology, psychosomatics, and semiotics. Gudrun von Uexküll was Jakob's wife.

3. The following reading draws in large part on Timothy Lenoir's excellent history, *The Strategy of Life: Teleology and Mechanics in Nineteenth Century German Biology* (Dordrecht, Holland: D. Reidel, 1982), and Stephen Jay Gould's *Ontogeny and Phylogeny* (Cambridge: Harvard University Press, 1977).

4. Stanley E. Salthe, *Development and Evolution: Complexity and Change in Biology* (Cambridge: MIT Press, 1993), 56–57. Deleuze addresses Baer's law in *Difference and Repetition* (trans. Paul Patton [New York: Columbia University Press, 1994]), which we examine in chapter 5.

5. Karl Lorenz, "Methods of approach to the problems of behavior," *Studies in Animal and Human Behavior (Volume II)*, trans. Robert Martin (Cambridge: Harvard University Press, 1971), 274.

6. Thomas S. Kuhn, *The Structure of Scientific Revolution*, 3rd edition (Chicago: University of Chicago Press, 1996).

7. On this note I find myself in agreement with T. Arthur Thomson, who claimed the following in an early review of Uexküll's book in 1927: "it is disappointing to have a thinker of von Uexküll's caliber repeating the libel after Darwin" (417), and, in reference to Uexküll's claim that Darwinism "is a religion rather than a science" (TB, 264), "If ever prejudice spoke it is here!" (418).

8. While many of Uexküll's criticisms may be deemed wrong, his view that Darwinism emphasizes mechanical interactions applies in a different manner to some Darwinians today. For example, Richard Dawkins, a leading Darwinian scholar, holds that organisms are nothing but "survival machines" for the true source of evolution: genetic reproduction (*The Selfish Gene*, 19 passim).

9. Beyond the occasional references, Uexküll devoted two very short writings exclusively to Kant: "Kant als Naturforscher. Von Erich Adikes." *Deutsche Rundschau*

53, 5 (1924): 209–10; and "Kants Einfluß auf die heutige Wissenschaft: Der große Königsberger Philosoph ist in der Biologie wieder lebendig geworden." *Preußische Zeitung* 9, 43 (1939): 3.

10. The passage that I am thinking of is found in Heidegger's 1927 lecture course, *The Basic Problems of Phenomenology* (trans. Albert Hofstadter. Bloomington: Indiana University Press, 1988), where he notes: "Self and world are not two beings, like subject and object, or like I and thou, but self and world are the basic determination of Dasein itself in the unity of the structure of being-in-the-world" (GA24, 422/297). The similarity, which is only superficial at this point, is in the necessity to see both subject/self and world as intrinsically related. However, Uexküll's subject is not the same as Heidegger's self, nor is the concept of world the same. I shall return to the similarities and differences between Uexküll and Heidegger in the next chapter.

11. Harald Lassen, "Leibniz'sche Gedanken in der Uexküll'schen Umweltlehre," in *Acta Biotheoretica* A5 (1939): 41–50.

12. A similar example of an animal's unusually persistent life can be found in Johnjoe McFadden's *Quantum Evolution: How Physics' Weirdest Theory Explains Life's Biggest Mystery* (New York: W.W. Norton, 2000), in which he explains the marvelous adaptability of a species of nematode worm that survives the dry and frozen Antarctic for decades, if not centuries, until a slight source of ice melts and thus sets off a rapid procedure of feasting and reproducing, before freezing over again (22).

13. See Alphonso Lingis's essays "Orchids and muscles," "Segmented organisms," "Beastiliatiy"; James Lovelock, *Gaia: A New Look at Life on Earth* (Oxford: Oxford University Press, 1979); and Lee Smolin, *The Life of the Cosmos* (Oxford: Oxford University Press, 1997).

14. For further reading on biology from a Hegelian standpoint, see Richard Levins and Richard Lewontin, *The Dialectical Biologist* (Cambridge: Harvard University Press, 1985).

15. Thomas A. Sebeok, *Perspectives in Zoosemiotics* (The Hague: Mouton, 1972).

16. McFadden, *Quantum Evolution*, chapter 3; and Richard Dawkins, *The Selfish Gene*, chapter 3. Though neither are biosemiotians, both describe the process of DNA replication as a form of communication.

17. Kalevi Kull, "Biosemiotics in the twentieth century: a view from biology," *Semiotica* 127, 1/4 (1999): 385–414, 386. On this history, see also Thomas Sebeok's *Global Semiotics* (Bloomington: Indiana University Press, 2001).

18. See Thure von Uexküll, "Introduction," 282; and Jesper Hoffmeyer's *Signs of Meaning in the Universe*, trans. Barbara J. Haveland (Bloomington: Indiana University Press, 1996), 56.

19. On Uexküll's influence and appearance in Cassirer's work, see Barend van Heusden's "Jakob von Uexküll and Ernst Cassirer," in *Semiotica* 134, 1/4 (2001): 275–92. On some of the relation Uexküll had with Cassirer, see Gudrun von Uexküll's biography.

20. Along with Hoffmeyer (1996) and Sebeok (1972 and 2001), see John Deely's *Basics of Semiotics* (Bloomington: Indiana University Press, 1990) and Robert S.

Corrington's *Ecstatic Naturalism: Signs of the World* (Bloomington: Indiana University Press, 1994), for some connection between Uexküll and semiotics.

21. For more on the repercussions of life beginning with two things, see H. Peter Steeves, "A Quantum Magical Realism Writ Small: Self-Referentiality, Information, and the Origin of Life" (unpublished manuscript).

22. Lynn Margulis. *Symbiotic Planet: A New Look at Evolution* (New York: Basic Books, 1998), 33.

CHAPTER 2: MARKING A PATH INTO THE ENVIRONMENTS OF ANIMALS

1. Karl Löwith, *Nature, History, and Existentialism*, ed. Arnold Levison (Evanston, IL: Northwestern University Press, 1966), 36.

2. See, among others, David Farrell Krell's *Daimon Life: Heidegger and Life Philosophy* (Bloomington: Indiana University Press, 1992), chapter one; Michel Haar's *The Song of the Earth: Heidegger and the Grounds of the History of Being*, trans. Reginald Lilly (Bloomington: Indiana University Press, 1993), chapter two; and Miguel de Beistegui's *Thinking with Heidegger: Displacements* (Bloomington: Indiana University Press, 2003), chapter one.

3. An interesting reading of Heidegger's (mis)treatment of material beings (entities, things, objects, tools), in and around *Being and Time*, can be found in Graham Harmon's *Tool-Being: Heidegger and the Metaphysics of Objects* (Chicago: Open Court, 2002), particularly chapter one.

4. This is a point that receives its greatest influence from Foucault's analyses of epistemes, especially by what he calls "The Anthropological Sleep" in *The Order of Things: An Archeology of the Human Sciences*, trans. unknown (New York: Vintage Books, 1994).

5. The theme of "metontology" is not one that I pursue here. On Heidegger and metontology, see Heidegger's comments in his 1928 lecture course, in which metontology is linked with an analysis of "beings as a whole" and the overturning of ontology: "As a result, we need a special problematic which has for its proper theme beings as a whole [*das Seiende im Ganzen*]. This new investigation resides in the essence of ontology itself and is the result of its overturning [*Umschlag*], its metabole" (GA26, 199/157). For a reading of Heidegger on metontology, see William McNeill's "Metaphysics, Fundamental Ontology, Metontology 1925–35," *Heidegger Studies* 8 (1992): 63–79; and Miguel de Beistegui's *Thinking with Heidegger* (chapter three).

6. There are many examples of texts devoted to either camp within the 1920s, as well as the surrounding decades. A useful reading of the development of "organicism" as another alternative to mechanism and vitalism in the early twentieth century is Donna Haraway's *Crystals, Fabrics, and Fields: Metaphors of Organicism in Twentieth-Century Developmental Biology* (New Haven: Yale University Press, 1976).

7. Emmanuel Levinas, "The Paradox of Morality: an Interview with Emmanuel Levinas," interview by Tamra Wright, Peter Hughes, and Alison Ainley, trans. Andrew Benjamin and Tamra Wright. *The Provocation of Levinas: Rethinking the Other*, ed. Robert Bernasconi and David Wood (New York: Routledge Press, 1988), 172.

8. Hans Jonas, *Mortality and Morality: A Search for the Good after Auschwitz*, ed. Lawrence Vogel (Evanston, IL: Northwestern University Press, 1996), 47.

9. Frank Schalow, *The Incarnality of Being: The Earth, Animals and the Body in Heidegger's Thought* (Albany: State University of New York Press, 2006).

10. F. J. J. Buytendijk, Zur Untersuchung des Wesensunterschieds von Mensch und Tier (in *Blätter für Deutsche Philosophie*, Vol. 3. Berlin: 1929–30).

11. Hans Driesch, *The History and Theory of Vitalism*, trans. C. K. Ogden (London: Macmillan, 1914), 2.

12. The reference to Uexküll is caught by Heinz Maeder, the student responsible for taking the 'minutes' of Heidegger's third class (GA85, 137). The seminar consists of a series of fragmentary notes, but among them one finds the reference to Uexküll.

13. For more on Heidegger's reading of *phusis*, see his 1939 essay "On the Essence and Concept of Φύσις in Aristotle's *Physics* B, I" (GA9). Just as he does with the concept of "world," in this essay he offers the Western history of "nature" with particular emphasis on Aristotle.

14. Heidegger's terminology is important here, particularly in consideration of how he will describe animal life, which we will take up below. For the moment, I want to highlight Dasein's "absorption [*Aufgehen*]," which must be seen in contrast to the "absorption [*Eingenommenheit*]" of animals. The equivocal term "absorption"does not speak to these two concepts in the same way.

15. One could easily imagine other areas where their similarities may be at issue. For example, one might raise the Aristotelian distinction between nourishment and sensation (*De Anima* II, 1–5). Or one might ask Bentham's question and wonder if plants have the capacity to suffer, as animals do (*The Principles of Morals and Legislation* [Amherst, NY: Prometheus Books, 1988], 311). For Heidegger, one might wonder if there isn't a distinction in how plants 'die' with respect to how animals 'die.' Likewise, the capacity for motility may be at issue as well. These issues may pertain not only to the plant/animal parallel, but also to the issue of whether all animals are essentially the same.

CHAPTER 3: DISRUPTIVE BEHAVIOR

1. Andy Clark, *Mindware: An Introduction to the Philosophy of Cognitive Science* (New York: Oxford University Press, 2001), 2.

2. On this reference, see Marjorie Grene, *Approaches to a Philosophical Biology* (New York: Basic Books, 1965), 70–71; and Rüdiger Safranski, *Martin Heidegger: Between Good and Evil*, trans. Ewald Osers (Cambridge: Harvard University Press, 1998), 198.

3. A similar statement is made in a contemporaneous essay, "Geschlecht II: Heidegger's Hand," trans. John P. Leavey, *Deconstruction and Philosophy*, ed. John Sallis (Chicago: University of Chicago Press, 1987), 173. This is a recurring theme in Derrida's oeuvre, as he reiterates this same problem in his later essay, "The Animal that Therefore I Am (More to Follow)," trans. David Wills, *Critical Inquiry* 28 (2002): 369–418. From the latter essay, take for example: "A critical uneasiness will persist;

in fact a bone of contention will be incessantly repeated throughout everything that I wish to develop. It would be aimed in the first place, once again, at the usage, in the singular, of a notion as general as 'the Animal,' as if all nonhuman living things could be grouped without the common sense of this 'commonplace,' the Animal, whatever the abyssal differences and structural limits that separate, in the very essence of their being, all 'animals,' a name that we would therefore be advised, to begin with, to keep within quotation marks" (402). In this same essay, which is just one of a projected number of essays on animals and various philosophers, Derrida notes that he will be giving a paper dealing with just Heidegger's 1929–1930 lecture course. But this essay does not appear in the original conference proceedings, *L'animal autobiographique: autour de Jacques Derrida*, ed. Marie-Louise Mallet (Paris: Galilée, 1999), nor has it, to the best of my knowledge, appeared in any collections since. The theme of the singular or lone animal ("the animal") versus the plurality of animals (either animals as such, or a group of animals) is also given attention by David Morris in "Animals and humans, thinking and nature," *Phenomenology and the Cognitive Sciences* 4 (2005): 49–72.

4. On this point Heidegger would differ somewhat with Irene Klaver's position in her essay "Stone Worlds: Phenomenology on (the) Rocks," *Eco-Phenomenology: Back to the Earth Itself*, eds. Charles S. Brown and Ted Toadvine (Albany: State University of New York Press, 2003). It is not a disagreement so much as an issue of interpretation in the use of "stone worlds." Stone worlds, in her account, appear when "the stone was encountered and taken up in a world" (357). Though this is not a Heideggerian analysis, it clearly runs counter to the thesis I have just recounted. Klaver's essay, however, presents an interesting reading of stones particularly in relation to Merleau-Ponty's philosophy of nature.

5. For a more comprehensive treatment on the differences between organs and equipment, and its importance to the 1929–1930 course, see Marc Richir's *Phenomenologie et Institution Symbolique* (Pairs: Éditions Jerome Millon, 1988); David Krell's *Daimon Life: Heidegger and Life-Philosophy* (Bloomington: Indiana University Press, 1992); William McNeill's "Life Beyond the Organism: Animal Being in Heidegger's Freiburg Lectures, 1929–30," *Animal Others: On Ethics, Ontology, and Animal Life*, ed. H. Peter Steeves (Albany: State University of New York Press, 1999); reprinted in *The Time of Life: Heidegger and Êthos*; and Matthew Calarco's "Heidegger's Zoontology," *Animal Philosophy: Essential Readings in Continental Thought*, ed. Matthew Calarco and Peter Atterton (New York: Continuum, 2004). I am indebted to each of these careful exegeses of Heidegger's lecture course and would refer the reader to them for further consultation on many discrepancies that I pass over in order to focus my attention on the ontological status of animal behavior. On a side note, Richir interestingly follows his reading of the 1929–1930 course with a section on animal ethology, in which he takes up the thought of Konrad Lorenz.

6. These etymological links are also noted by the English translators, William McNeill and Nicholas Walker, in their "Foreword" and notes to Heidegger's *The Fundamental Concepts of Metaphysics*.

7. Heidegger, GA 29/30, 336/230. Following the previous note on the ape's not having "hands," doesn't it seem funny that Heidegger has no problem attributing "eyes" to the bee, even if "the bee's eye has neither pupil, nor iris, nor lens"?

Admittedly, he does put "eye" once in single quotes: '*Auge.*' It is also the case that it is the capacity, rather than the organ, that interests Heidegger. Nevertheless, it is a curiosity that apes do not have hands, but "organs that grasp," whereas bees have eyes, and not simply visual organs.

8. In Plato's *The Republic*, trans. Paul Shorey (Cambridge: Harvard University Press, 1969), 453c–e, Socrates compares his situation in the argument with one who is lost at sea. What can one do but swim? This is the only way out of the aporia: to keep going, keep talking. In *Protagoras*, it is the opposite: rather than finding one's way home, the whole dialogue ends in an "extraordinary tangle" (361c). See Sarah Kofman on this theme in her *Comment s'en sortir?* (Paris: Galilée, 1983).

9. For an ethical reading of Heidegger and the question of "speaking for" animals, which I am not pursuing here, see Frank Schalow's "Who Speaks for the Animals? Heidegger and the Question of Animal Welfare," *Environmental Ethics* 22 (2000): 259–71. Reprinted in *The Incarnality of Being*.

10. On "leeway," which only seldomly reappears, see 384/264, and even more interestingly, 497/342, where a comparable leeway is associated with human freedom: "This provision of, and subjection to, something binding is in turn only possible where there is *freedom*. Only where there is this possibility of transferring our being bound from one thing to another are we given the leeway [*Spielraum*] to decide concerning the conformity or non-conformity of our comportment to whatever is binding." Leeway, in the case of human freedom, is precisely that which signifies that we are *not* bound, that our relation is one of *comportment*.

11. John van Buren, *The Young Heidegger: Rumour of the Hidden King* (Bloomington: Indiana University Press, 1994), 359. David Krell (*Daimon Life*, 90) has also drawn attention to this passage in order to emphasize how the concept of "life" is applied to both snail and Dasein.

12. I am indebted to John van Buren for bringing this passage to my attention. The passage is taken from Heidegger's typed manuscript of a lecture he delivered on May 24, 1926, "Vom Wesen der Wahrheit." The manuscript reads: "*Auch eine Qualle hat schon, wenn sie ist, ihre Welt. So etwas wie Welt, Seiendes, das sie nicht selbst ist, ist ihr entdeckt, aufgeschlossen*" (8). Since the submission of this manuscript, Heidegger's lecture "On the Essence of Truth" has been published for the first time in Theodore Kisiel and Thomas Sheehan (eds.), *Becoming Heidegger: On the Trail of His Early Occasional Writings, 1910–1927* (Evanston: Northwestern University Press, 2007). The passage reads: "Even a jellyfish already has, when it is, its world. Something like a world, a being that it itself is not, is revealed to it, uncovered" (284).

13. On Heidegger's earlier usage of *Umwelt* in the late teens and early twenties, see "Appendix D" to Theodore Kisiel's *The Genesis of Heidegger's Being and Time* (Berkeley: University of California Press, 1993), 506.

14. In a different context from the present concern, David Krell looks at the political ramifications of various etymological cognates of *Erschütterung* throughout Heidegger's writings of the 1930s. Chapter five of *Daimon Life* offers an etymology of *Scheitern, Scheiden, Erschütterung, Schütten,* and *Schütteln*.

15. For a sustained treatment of Heidegger on the *Augenblick*, see William McNeill's *The Glance of the Eye: Heidegger, Aristotle, and the Ends of Theory* (Albany:

State University of New York Press, 1999), and in particular the subchapter "The Time of the Augenblick."

16. Françoise Dastur, *Heidegger et la question anthropologique* (Louvain-Paris: Éditions Peeters, 2003), 62.

17. On touch (*Rührung, beruhren*), recall his notes on the lizard (GA 29/30, 290–91/196–97). On stimulation (*Reiz, Reizen*), see ibid., 372–73/256.

18. Derrida, "The Animal that Therefore I Am," 391. Derrida notes many times throughout this essay that he will have to return to a more careful reading of Heidegger and animals, just as he does here with respect to this passage on time. Sadly, however, he may have been right when he states: "(my hypothesis is this: whatever is put off until later will probably be put off for ever; later here signifies never)," 391. I know of no sustained reading by Derrida on this lecture course outside of the present essay and brief remarks here and there (e.g., *Of Spirit*; " 'Eating Well,' or the Calculation of the Subject," trans. Peter Connor and Avital Ronnell, *Points . . . Interviews 1974–94*, ed. Elisabeth Weber [Stanford: Stanford University Press, 1995]).

19. McNeill, "Life Beyond the Organism," 239. This essay has been reprinted in William McNeill's *The Time of Life: Heidegger and Êthos* (Albany: State University of New York Press, 2006).

20. This portion of the passage is cited from David Krell's *Daimon Life* (26).

CHAPTER 4: THE THEME OF THE ANIMAL MELODY

1. As just one explicit example, see Douglas Low's "The continuing relevance of *The Structure of Behavior*," *International Philosophical Quarterly* 44, 3 (2004): 411–30.

2. Many have worked on these notes, but, as an explicit example, see Douglas Low's *Merleau-Ponty's Last Vision: A Proposal for the Completion of* The Visible and the Invisible (Evanston, IL: Northwestern University Press, 2000).

3. Although I am largely sympathetic to Mark B. N. Hansen's reading—whose essay has many similarities to my own project—I don't agree with his claim that there is an "ontological turn," or "failure," in the early writings. See "The Embryology of the (In)visible," in *The Cambridge Companion to Merleau-Ponty*, eds. Taylor Carman and Mark B. N. Hansen (Cambridge: Cambridge University Press, 2005).

4. Renaud Barbaras, *The Being of the Phenomenon: Merleau-Ponty's Ontology*, trans. Ted Toadvine and Leonard Lawlor (Bloomington: Indiana University Press, 2004). M. C. Dillon, *Merleau-Ponty's Ontology* (Evanston, IL: Northwestern University Press, 1997), 85.

5. For a particular look at the concept of "archeology" (and *arche*) and its application to Merleau-Ponty's thought, see Leonard Lawlor's reading in *Thinking Through French Philosophy: The Being of the Question* (Bloomington: Indiana University Press, 2003), chapter 2. Lawlor addresses not only Merleau-Ponty, but Freud, Husserl, Kant, and, more specifically, Foucault.

6. I noted this earlier, but it bears repeating: "behavior" is the translation of Merleau-Ponty's "*comportement*." This distinction is highly important when compared with Heidegger's terminology, where the English translation of *Benehmen* (behavior) refers specifically to animal behavior and the English translation of *Verhaltung* (comportment) refers specifically to human 'behavior.' As we will see, Merleau-Ponty does differentiate between the behavior of animals and humans, but he does not do so conceptually. One shouldn't confuse, therefore, Merleau-Ponty's use of "*comportement* [behavior]" with Heidegger's "behavior" and/or "comportment."

7. In keeping with Merleau-Ponty's wish to "start 'from below' " with behavior, we can note how "*il se creuse*" might be more literally translated as "hollowed out," "breaks apart," or even "dug up" as opposed to the more reader-friendly "opened up." I note this distinction for two reasons: first, as mentioned before, there is a faint archeological theme already at play in *The Structure of Behavior*, and we can read this in how behavior digs away at the natural world; second, as noted in the chapters on Heidegger, the English translation of "to open" carries a strong connotation in Heidegger's thought, whereas Merleau-Ponty is not thinking of *das Offene* (*l'ouvert*).

8. Renaud Barbaras, "The Movement of The Living as the Originary Foundation of Perceptual Intentionality," trans. Charles Wolfe. *Naturalizing Phenomenology: Issues in Contemporary Phenomenology and Cognitive Science*, eds. Jean Petitot et al. (Stanford: Stanford University Press, 1999), 532.

9. Kurt Goldstein, *The Organism: A Holistic Approach to Biology Derived from Pathological Data in Man* (New York: Zone Books, 2000), 18. This is Goldstein's own self-depiction.

10. As far as I know, the biographical answer to Merleau-Ponty's usage of Uexküll is undocumented. However, just to hazard another textual guess, Goldstein may again appear to be the common denominator since he also pays lip service to Buytendijk in a manner that would have been meaningful to Merleau-Ponty: "As we have seen, animal behavior cannot be understood as a summation of single processes. It points, rather, to an individual organization. . . ." In this general characteristic of animal nature, I find myself in agreement with Alverdes, Frederik Buytendijk, and others. Likewise for the animal, the environment is not given as absolute but arises in the animal's being and acting" (*The Organism*, 355). To hazard a guess, Merleau-Ponty was probably drawn to Uexküll via Buytendijk via Goldstein. Georges Canguilhem may also have had a role in this. (I am thankful to Charles Wolfe for bringing this to my attention.) As for his later and more explicit treatment of Uexküll in *Nature*, there is a more likely scenario: Uexküll's *A Stroll Through the Environment of Animals and Humans* and *The Theory of Meaning* appeared together in French translation in 1956 (*Mondes Animaux et Monde Humain*) just a year before Merleau-Ponty offers his lectures on him. It is this collection that Deleuze will also cite.

11. The historical relations between Heidegger, Merleau-Ponty, and Deleuze in their textual references to Uexküll and to each other are certainly interesting, but I'm afraid I don't have much to contribute beyond speculations I have already engaged in. For example, it is nearly certain that Merleau-Ponty did not have access to Heidegger's work in the 1929–1930 course. But while the publication of

Heidegger's lecture notes wouldn't appear until the 1980s, this does not rule out the possibility that Merleau-Ponty might have known of the lecture courses and their content. Likewise, it is possible and even likely that Deleuze knew of Merleau-Ponty's references to Uexküll from his 1957–1958 course.

12. Merleau-Ponty will recognize these terms as Uexküll's by the time of his *Nature* lectures (cf. 227/172).

13. Barbaras argues along these lines in many of his works, including *The Being of the Phenomenon* and "Merleau-Ponty and Nature."

14. He falls more in line with Leibniz in this respect (cf. Deleuze, *The Fold: Leibniz and the Baroque*, 9) than he does, for instance, with Philo's hyperbolic remarks about the world *as* animal in David Hume's *Dialogues Concerning Natural Religion* (Part VII).

15. Hansen, "The Embryology of the (In)visible," 252.

16. Glen Mazis, "Merleau-Ponty's Concept of Nature: Passage, The Oneiric, and Interanimality," *Chiasmi International. Merleau-Ponty: From Nature to Ontology* (VRIN, Mimesis, University of Mempthis Press, 2000), 232.

17. Alphonso Lingis, "The World as a Whole," *From Phenomenology to Thought, Errancy, and Desire: Essays in Honor of William J. Richardson, S.J.*, ed. Babette Babich (Dordrecht: Kluwer Academic Publishers, 1995).

18. The material that constitutes the third lecture course on nature is a series of Merleau-Ponty's "sketches" (eight in all), in which he retreats similar key ideas in different ways. It is therefore the case that many of the sketches follow a similar order, though repeated differently each time. The association between Uexküll's *Umwelt* and the flesh is found at the beginning of the first sketch, but can also be found at the beginning of the second and third sketches as well.

19. On the issue of an "ontology of nature," see John Russon's essay "Embodiment and responsibility: Merleau-Ponty and the ontology of nature," *Man and World* 27 (1994): 291–308. Russon draws primarily from *Phenomenology of Perception* as well as *The Visible and the Invisible* (the *Nature* lectures were not yet available at this time), so while he mentions the concept of the *Umwelt*, there is no connection made to Uexküll.

20. Barbaras, "Merleau-Ponty and Nature," 37.

21. For an account of Merleau-Ponty's usage of "sense" in particular, and its application to an ontology of nature, see Ted Toadvine's "Singing the World in a New Key: Merleau-Ponty and the Ontology of Sense," *Janus Head* 7, 2 (2004): 273–83.

22. This example, which Merleau-Ponty reinterprets from Husserl's *Ideas II*, can be found in Merleau-Ponty's "The Philosopher and His Shadow" (*Signs*, trans. Richard C. McCleary [Evanton, IL: Northwestern University Press, 1964]), the third course on *Nature*, and the more well-known passages in *The Visible and the Invisible*.

23. For example, compare the remark on the rehabilitation of the sensible world in *Nature* to his remark in "The Philosopher and His Shadow" (S, 210/166–67) in which "the ontological rehabilitation of the sensible" is noted in the contact of the touching hand.

24. See also: "The relation between the circularities (my body-the sensible) does not present the difficulties that the relation between 'layers' or linear orders presents (nor the immanence-transcendence alternative)" (VI, 321/268).

25. Compare the following claim with Heidegger: "Reciprocally, human being is not animality (in the sense of mechanism) + reason.—And this is why we are concerned with the body: before being reason, humanity is another corporeity" (N, 269/208). The first sentence is quite similar to Heidegger (cf. GA2, 50/75) whereas the second sentence suggests a new direction. A more obvious site for further study on their relation, which I cannot undertake here, is Merleau-Ponty's lectures on Heidegger in the first part of *Notes des cours au Collège de France: 1957–58 et 1960–61* (Paris: Gallimard, 1996). Françoise Dastur addresses these lectures in *Chair et language* (Fougères: Encre Marine, 2001).

CHAPTER 5: THE-ANIMAL-STALKS-AT-FIVE-O'CLOCK

1. Following the convention of many other commentators, I adopt the practice of referring to the ontology in Deleuze and Guattari's collaborative writings as "Deleuze's ontology." Though this practice may not give Guattari his share of the credit, most of the ontological ideas that appear in their collaborations can be traced back to Deleuze's earlier writings, such as *Difference and Repetition*, *The Logic of Sense*, and his engagements with the philosophical tradition.

2. Another recent publication that takes up the theme of the 'being of the question,' and the question of being, in light of contemporary thought is Leonard Lawlor's *Thinking Through French Philosophy: The Being of the Question*. We might also cite one of Derrida's texts on Heidegger, in which the subtitle simply attests to the importance of the question as such: *Of Spirit: Heidegger and the Question*.

3. Deleuze invokes both Heidegger and Merleau-Ponty, albeit briefly, in his theorization of the problem and the question for his own ontology of difference (DR, 189–90/64). He will note: "We regard as fundamental this 'correspondence' between difference and questioning, between ontological difference and the being of the question" (DR, 91/66).

4. Bonta and Protevi, *Deleuze and Geophilosophy: A Guide and Glossary* (Edinburgh: Edinburgh University Press, 2004), vii–viii. As an overall evaluation of the impact of Deleuze's thought, we might recall Michel Foucault's infamous declaration that "perhaps one day, this century will be known as Deleuzian" (Michel Foucault, "Theatrum Philosophicum," in *Language, Counter-memory, Practice*, ed. and trans. Donald F. Boucher [Ithaca: Cornell University Press, 1977], 165). Foucault wrote this at the beginning of his review of Deleuze's two books, *Difference and Repetition* and *The Logic of Sense*, which establish his ontological outlook.

5. Miguel de Beistegui, *Truth and Genesis: Philosophy as Differential Ontology* (Bloomington: Indiana University Press, 2004), 190.

6. Keith Ansell Pearson, *Germinal Life: The Difference and Repetition of Deleuze* (New York: Routledge Press, 1999).

7. The bulk of Deleuze's lectures are not yet available in print, but they are steadily appearing on the internet in French, as well as the occasional Spanish and English translation: www.webdeleuze.com.

8. This essay can be found in Deleuze's *Spinoza: Practical Philosophy*, trans. Robert Hurley (San Francisco: City Lights Books, 1988). The first edition of this

book was published in 1970 as *Spinoza: textes choisis*, and later expanded for republication as *Spinoza: Philosophie practique* (Paris: Les Éditions de Minuit) in 1981. The essay "Spinoza and Us," included in the second edition, was first published in 1978 in *Revue de Synthèse III*: 89–91.

9. De Landa, *Intensive Science and Virtual Philosophy*, 82. Cf., DR, 299–300/232–33.

10. De Landa explains in *Intensive Science and Virtual Philosophy* how Deleuze's distinction between the extensive and intensive actually draws directly from thermodynamics. Other sources that give a picture of the sciences underlying Deleuze's ontology include De Beistegui's *Truth and Genesis* (chapter 6, "Physics beyond Metaphysics?"), Bonta and Protevi's *Deleuze and Geophilosophy*, and Timothy S. Murphy, "Quantum Ontology: A Virtual Mechanics of Becoming," in *Deleuze and Guattari: New Mappings in Politics, Philosophy, and Culture*, eds. Eleanor Kaufman and Kevin Jon Heller (Minneapolis: University of Minnesota Press, 1998).

11. Keith Ansell Pearson makes a similar point in *Germinal Life* (175).

12. De Beistegui, *Truth and Genesis*, 187; De Landa, *Intensive Science and Virtual Philosophy*, 4.

13. Timothy S. Murphy, "Quantum Ontology: A Virtual Mechanics of Becoming"; Constantin V. Boundas' "Deleuze-Bergson; an Ontology of the Virtual," in *Deleuze: A Critical Reader*, ed. Paul Patton (Oxford: Blackwell Press, 1996); and Constantin V. Boundas, "An Ontology of Intensities," *Epoché* 7, 1 (2002): 15–37, respectively.

14. While I am remaining within the parameters set by Deleuze in this description of univocity, it is worth noting that the story goes back to at least Aristotle, and that Heidegger gives an account of it in his 1931 course, *Aristotle's* Metaphysics Θ *1–3* (GA 33, specifically §§4–6).

15. For further reading on the sources to Deleuze's theory of univocity, see de Beistgui's chapter "The Renewal of Ontology" in *Truth and Genesis*, Michael Hardt's *Gilles Deleuze: An Apprenticeship in Philosophy* (Minneapolis: University of Minnesota Press, 1993), and Paul Bains's "Umwelten," *Semiotica* 134, 1/4 (2001): 137–67.

16. Frederick Copleston, *Medieval Philosophy* (New York: Doubleday, 1993), 354.

17. To match Deleuze's fourfold, we are reminded of Foucault's own critique of "the four similitudes" (convenience, emulation, analogy, and sympathy) in *The Order of Things*. These concepts are common themes for critical appraisal in contemporary continental thought.

18. This passage, cited from a lecture course Deleuze delivered in 1974, can be found in Paul Bains's very topical essay "Umwelten" (143). On a side note, Deleuze's claim that "the tick is God" is not immediately comparable to the statement that "God is a lobster" that Deleuze and Guattari make in A *Thousand Plateaus* (54/40). In the latter statement, Deleuze and Guattari are expressing the "double articulation" that occurs in "stratification," the compacting, thickening, or capturing of different particle flows into certain strata. While the two statements are not addressing the same issue, they are not unrelated either: both address different aspects of being.

19. Onto-hetero-genesis is how De Beistegui describes Deleuze's differential ontology in *Truth and Genesis*. Similarly, Constantin V. Boundas, in a wonderful essay,

describes this process of difference as a 'philosophical heterology' in "What Difference Does Deleuze's Difference Make?" (*Symposium: Canadian Journal of Continental Philosophy* 10, 1 (2006): 397).

20. In one of his more sustained reflections on Heidegger, Deleuze addresses Heidegger on this very issue of a philosophy of difference (DR, 89–91/64–66). Another reading occurs in Deleuze's appraisal of Foucault's "final rediscovery of Heidegger" (F, 115–21/107–14).

21. Alain Badiou writes: "The virtual is the very Being of beings, or we can even say that it is beings qua Being, for beings are but modalities of the One, and the One *is* the living production of its modes" (48). This would appear to be a direct correlation, though it is not clear that Badiou is using 'Being' here in its Heideggerian sense. This is particularly unclear because Badiou is otherwise explicit in how Deleuze differs from Heidegger in terms of the univocity of being (21–23). However, if a direct connection is unclear with Badiou, it cannot by missed when Constantin Boundas writes: "The relation between intensity and extension is much like the Heideggerean relation between the ontological and the ontic or the Levinasian relation between saying and the said" ("An Ontology of Intensities," 19). There may be something to this relation, but Boundas does not elaborate on why he thinks they are related. Without further support, I fail to see the similarity. While I do not wish to avoid this question, I will only address it tangentially through the remaining part of this chapter where I address Deleuze's theory of virtual intensities. It is my hope that this discussion will at least shed some light on why I do not think the virtual/intensity is as close to Heidegger's Being as has been suggested. As Badiou notes with respect to the proximity of Heidegger to Deleuze, "This is an extremely complex question" (21).

22. For a reading of various forms of human history from a partially Deleuzian perspective, see Manuel De Landa's *A Thousand Years of Nonlinear History* (New York: Zone Books, 1997). An interesting companion to this would be Robert Frodeman's *Geo-Logic* (Albany: State University of New York Press 2003), which explores similar geological issues but from an Heideggerian standpoint.

23. Boundas, "What Difference Does Deleuze's Difference Make?," 398. From a more 'scientific' perspective, Bonta and Protevi provide a useful description of how the virtual, intensive, and actual/extensive coalesce into a 'unified' ontological voice: "The structure of the virtual realm can be explicated as a meshed continuum of heterogeneous multiplicities defined by zones of indiscernability or 'lines of flight' [a move of deterritorialization, almost literally an escape from one's milieu in a new becoming]. The virtual realm contains the patterns and thresholds of behavior of the material systems in their intensive (far-from-equilibrium and near thresholds of self-order) and actual (equilibrium, steady state, or stability) conditions" (16).

24. Although I am not following the role Darwin occupies in Deleuze's writings, he contributes in at least two ways. In *Difference and Repetition*, Darwin (together with Freud) is said to introduce individual difference into the connective play of living relations, such as through natural selection. "Darwin's great novelty, perhaps, was that of inaugurating the thought of individual difference. The leitmotiv of *The Origin of Species* is: we do not know what individual difference is capable of!" (319/248). In *A Thousand Plateaus*, Darwin again receives accolades for a similar

contribution: "Darwinism's two fundamental contributions move in the direction of a science of multiplicities: the substitution of populations for types, and the substitution of rates or differential relations for degrees" (63–64/48). While Deleuze finds certain things problematic in the theory of evolution—such as genealogy, kinship, descent, filiation—Darwin is also recognized as having offered a new look at the production of difference.

25. De Landa, *A Thousand Years of Nonlinear History*, 104.

26. For example, he notes how "Diogenes Laertius relates that the Stoics compared philosophy to an egg," which leads him to ask: "Is not Humpty Dumpty himself the Stoic master?" (LS, 167/142). The details of this comparison matter less for our interests than the general observation that the egg is associated with both Humpty Dumpty (perhaps the most famous egg in literature) and Deleuze's beloved Stoics.

27. On the changing nature of Deleuze's concepts, De Landa notes: "Gilles Deleuze changes his terminology in every one of his books. Very few of his concepts retain their names or linguistic identity. The point of this terminological exuberance is not merely to give the impression of difference through the use of synonyms, but rather to develop a set of *different* theories on the *same* subject, theories which are slightly displaced relative to one another but retain enough overlaps that they can be meshed together as a heterogeneous assemblage" (*Intensive Science and Virtual Philosophy*, 202). Just as De Landa has done, I have sacrificed some consistency in my usage of Deleuze's fluctuating terminology for the sake of portraying a clearer picture of what is at work in his ontology of organisms. De Landa's "Appendix: Deleuze's Words," as well as Bonta and Protevi's *Deleuze and Geophilosophy*, provide good sources for tracing the various connections between different concepts.

28. De Landa, *A Thousand Years of Nonlinear History*, 34.

29. Richard Dawkins and Johnjoe McFadden have each described how biologists and chemists have attempted to reproduce conditions that might approximate the origins of life some three to four thousand million years ago. The success of these experiments, however, has not yet rivalled the pure fascination that such work inspires. We should also recall Heidegger's reference to the "primeval slime" (*Urschleim*) that he raises in his dismissal of Darwin's developmental theory in the 1929–1930 course (GA29/30, 402/277).

30. "The Autonomy of Affect," *Deleuze: A Critical Reader*, ed. Paul Patton (Oxford: Blackwell Press, 1996), 231.

31. De Landa, *Intensive Science and Virtual Philosophy*, 58.

32. McFadden, *Quantum Evolution*, 133; cf. 92–95. James Gleick, in his book *Chaos* (New York: Penguin, 1987), also explains: "The concept of entropy comes from thermodynamics, where it serves as an adjunct of the Second Law, the inexorable tendency of the universe, and any isolated system in it, to slide toward a state of increasing disorder. . . . Entropy is the name for the quality of systems that increases under the Second Law—mixing, disorder, randomness" (257).

33. Chaosmos is the term used to designate the in-between of rhythm and chaos, the "rhythm-chaos," the disjunctive conjunction of chaos and cosmos. Fred Evans explains that the term derives from James Joyce's *Finnegans Wake*. Evans writes: "He draws the view that everything is penetrated with everything else, at

once cosmos and chaos" ("Chaosmos and Merleau-Ponty's View of Nature," *Chiasmi International 2, From Nature to Ontology* [Paris: VRIN, Mimesis; Memphis: University of Memphis Press, 2000], 72).

34. One similarly reads how "the ideally continuous belt or ring of the stratum . . . exists only as shattered, fragmented into epistrata and parastrata" (ATP, 68–69/52).

35. Ronald Bogue, *Deleuze on Music, Painting, and the Arts* (New York: Routledge, 2003), 18.

36. On the orchid and the wasp, I would also recommend Alphonso Lingis's essays "Orchids and Muscles" (*Exceedingly Nietzsche*, eds. David Farrell Krell and David Wood [New York: Routledge, 1988]) and "Bestiality" (*Animal Others*, ed. H. Peter Steeves [Albany: State University of New York Press, 1999]). The latter in particular offers an illustrative account of the symbiotic relationship between Brazil nut flowers and wasps. Lingis clearly draws as much from Deleuze and Guattari as he does from the phenomenological tradition.

37. Arnaud Villani, *La guêpe et l'orchidée: Essai sur Gilles Deleuze* (Paris: Belin, 1999).

38. Moira Gatens draws a helpful distinction between biology and ethology: "biology lays down rules and norms of behavior and action, whereas ethology does not claim to know, in advance, what a body is capable of doing or becoming," "Through a Spinozist Lens: Ethology, Difference, Power," *Deleuze: A Critical Reader*, ed. Paul Patton (Oxford: Blackwell Press, 1996), 169.

39. Boundas, "What Difference Does Deleuze's Difference Make?," 399.

Bibliography

Agamben, Giorgio. *The Coming Community*. Translated by Michael Hardt. Minneapolis: University of Minnesota Press, 1993.

———. *The Open: Man and Animal*. Translated by Kevin Attell. Stanford: Stanford University Press, 2004.

Alliez, Eric. *The Signature of the World, Or, What Is Deleuze and Guattari's Philosophy?* Translated by Eliot Ross Albert and Alberto Toscano. New York: Continuum Press, 2004.

Ansell Pearson, Keith. *Germinal Life: The Difference and Repetition of Deleuze*. London: Routledge Press, 1999.

Aristotle. *History of Animals, Books VII–X*. Translated by D. M. Balme. Cambridge: Harvard University Press, 1991.

———. *On the Soul*. Translated by H. S. Hett. Cambridge: Harvard University Press, 1995.

Babich, Babette, Debra Bergoffen, and Simon Glynn, eds. *Continental and Postmodern Perspectives in the Philosophy of Science*. New York: Aveburg Press, 1995.

Badiou, Alain. *Deleuze: The Clamor of Being*. Translated by Louise Burchill. Minneapolis: University of Minnesota Press, 2000.

Bains, Paul. "Umwelten." *Semiotica* 134, 1/4 (2001): 137–167.

Barbaras, Renaud. "The Movement of The Living as The Originary Foundation of Perceptual Intentionality." Translated by Charles Wolfe. *Naturalizing Phenomenology: Issues in Contemporary Phenomenology and Cognitive Science*, eds., Jean Petitot, Francisco J. Varela, Bernard Pachoud, and Jean-Michel Roy. Stanford: Stanford University Press, 1999.

———. "Merleau-Ponty and Nature." Translated by Paul Milan, *Research in Phenomenology* 31 (2001): 22–38.

———. *Vie et intentionnalité: Recherches phénoménologiques*. Paris: J. Vrin, 2003.

———. *On the Being of the Phenomenon*. Translated by Ted Toadvine and Leonard Lawlor. Bloomington: Indiana University Press, 2004.

Beistegui, Manuel de. *Thinking with Heidegger: Displacements*. Bloomington: Indiana University Press, 2003.

———. *Truth and Genesis: Philosophy as Differential Ontology*. Bloomington: Indiana University Press, 2004.

Bentham, Jeremy. *The Principles of Morals and Legislation*. Amherst, NY: Prometheus Books, 1988.

Bergson, Henri. *Creative Evolution*. Translated by Arthur Mitchell. New York: Dover, 1998.
Bogue, Ronald. *Deleuze on Music, Painting, and the Arts*. New York: Routledge Press, 2003.
Bonta, Mark, and John Protevi. *Deleuze and Geophilosophy: A Guide and Glossary*. Edinburgh: Edinburgh University Press, 2004.
Boundas, Constantin V. "Deleuze-Bergson; an Ontology of the Virtual." *Deleuze: A Critical Reader*, ed., Paul Patton. Oxford: Blackwell, 1996.
———. "An Ontology of Intensities." *Epoché* 7, 1 (2002): 15–37.
———. "What Difference Does Deleuze's Difference Make?" *Symposium: Canadian Journal of Continental Philosophy* 10, 1 (2006): 397–423.
Brown, Charles S., and Ted Toadvine, eds. *Eco-Phenomenology: Back to the Earth Itself*. Albany: State University of New York Press, 2003.
Buren, John van. *The Young Heidegger: Rumor of the Hidden King*. Bloomington: Indiana University Press, 1994.
Carbone, Mauro. *The Thinking of the Sensible: Merleau-Ponty's A-Philosophy*. Evanston, IL: Northwestern University Press, 2004.
Calarco, Matthew. "Heidegger's Zoontologies." *Animal Philosophy: Essential Readings in Continental Thought*, eds. Matthew Calarco and Peter Atterton. New York: Continuum, 2004.
Canguilhem, Georges. *La connaissance de la vie*. Paris: J. Vrin, 1969.
Cassirer, Ernst. *An Essay on Man: An Introduction to a Philosophy of Human Culture*. New Haven: Yale University Press, 1972.
———. *The Problem of Knowledge: Philosophy, Science, and History Since Hegel*. Translated by William H. Woglom and Charles W. Hendel. New Haven: Yale University Press, 1978.
Clark, Andrew. *Being There: Putting Brain, Body, and World Together Again*. Cambridge: MIT Press, 1997.
———. *Mindware: An Introduction to the Philosophy of Cognitive Science*. Oxford: Oxford University Press, 2001.
Copleston, Frederick. *Medieval Philosophy*. New York: Doubleday, 1993.
Corrington, Robert S. *Ecstatic Naturalism: Signs of the World*. Bloomington: Indiana University Press, 1994.
Dastur, Françoise. *Chair et language*. Fougères: Encre Marine, 2001.
———. *Heidegger et la question anthropologique*. Louvain-Paris: Éditions Peeters, 2003.
Dawkins, Richard. *The Selfish Gene*. Oxford: Oxford University Press, 1989.
Deely, John. *Basics of Semiotics*. Bloomington: Indiana University Press, 1990.
Delanda, Manuel. *A Thousand Years of Nonlinear History*. New York: Zone Books, 1997.
———. *Intensive Science and Virtual Philosophy*. London: Continuum Press, 2002.
Deleuze, Gilles. *Différence et Répétition*. Paris: Presses Universitaires de France, 1968. (*Difference and Repetition*. Translated by Paul Patton. New York: Columbia University Press, 1994.)
———. *Logique du Sens*. Paris: Les Éditions de Minuit, 1969. (*The Logic of Sense*. Edited by Constantin Boundas. Translated by Mark Lester and Charles Stivale. New York: Columbia University Press, 1990.)

———. *Spinoza et le problème de l'expression*. Paris: Éditions de Minuit, 1968. (*Expressionism in Philosophy: Spinoza*. Translated by Martin Joughin. New York: Zone Books, 1992.)

———. *Spinoza: Philosophie practique*. Paris: Les Éditions de Minuit, 1981. (*Spinoza: Practical Philosophy*. Translated by Robert Hurley. San Francisco: City Lights Books, 1988.)

———. *Foucault*. Paris: Éditions de Minuit, 1986. (*Foucault*. Translated by Seán Hand. Minneapolis: University of Minnesota Press, 1988.)

———. *The Fold: Leibniz and the Baroque*. Translated by Tom Conley. Minneapolis: University of Minnesota Press, 1993.

Deleuze, Gilles, and Félix Guattari. *Mille Plateaux*. Paris: Les Éditions de Minuit, 1980. (*A Thousand Plateaus*. Translated by Brian Massumi. Minneapolis, University of Minnesota Press, 1987.)

———. *Qu'est-ce que la philosophie?* Paris: Éditions de Minuit, 1991. (*What Is Philosophy?* Translated by Hugh Tomlinson and Graham Burchell. New York: Columbia University Press, 1994.)

Dennett, Daniel. *Darwin's Dangerous Idea: Evolution and the Meanings of Life*. London: Allen Lane, 1995.

Derrida, Jacques. *Of Spirit: Heidegger and the Question*. Translated by Geoff Bennington and Rachel Bowby. Chicago: University of Chicago Press, 1987.

———. "Geschlecht II: Heidegger's Hand." Translated by John P. Leavey, Jr. *Deconstruction and Philosophy*. Edited by John Sallis. Chicago: University of Chicago Press, 1987.

———. "'Eating Well,' or the Calculation of the Subject." Translated by Peter Connor and Avital Ronnell. *Points . . . Interviews 1974–94*. Edited by Elisabeth Weber. Stanford: Stanford University Press, 1995.

———. "The Animal That Therefore I Am (More to Follow)." Translated by David Wills. *Critical Inquiry* 28 (2002): 369–418.

Dillon, M. C. *Merleau-Ponty's Ontology*. Evanston, IL: Northwestern University Press, 1997.

Driesch, Hans. *The History and Theory of Vitalism*. Translated by C. K. Ogden. London: Macmillan, 1914.

Elden, Stuart. "Heidegger's Animals." *Continental Philosophy Review* 39 (2006): 273–291.

Evans, Fred. "Chaosmos and Merleau-Ponty's View of Nature." *Chiasmi International* 2, *From Nature to Ontology*. Paris: VRIN, Mimesis; Memphis: University of Memphis Press, 2000.

Evans, Fred, and Leonard Lawlor, eds. *Chiasms: Merleau-Ponty's Notion of Flesh*. Albany: State University of New York Press, 2000.

Evernden, Neil. *The Natural Alien: Humankind and Environment*. Toronto: University of Toronto Press, 1985.

Foucault, Michel. "Theatrum Philosophicum." *Language, Counter-memory, Practice*. Edited and Translated by Donald F. Boucher. Ithaca: Cornell University Press, 1977.

———. "Introduction" to Georges Canguilhem, *The Normal and the Pathological*. Translated by Carolyn R. Fawcett. New York: Zone Books, 1991.

———. *The Order of Things: An Archaeology of the Human Sciences*. New York: Vintage Books, 1994.

Franck, Didier. "Being and the Living." Translated by Peter T. Connor. *Who Comes After the Subject?* Edited by Eduardo Cadava, Peter Connor, and Jean-Luc Nancy. New York: Routledge Press, 1991.

Frodeman, Robert. *Geo-Logic: Breaking Ground Between Philosophy and the Earth Sciences*. Albany: State University of New York Press, 2003.

Gadamer, Hans-Georg. *Truth and Method*. Translated by Joel Weinsheimer and Donald G. Marshall. New York: Continuum Press, 1996.

Gatens, Moira. "Through a Spinozist Lens: Ethology, Difference, Power." *Deleuze: A Critical Reader*. Edited by Paul Patton. Oxford: Blackwell, 1996.

Gleick, James. *Chaos: Making a New Science*. New York: Penguin Books, 1987.

Glendinning, Simon. "Heidegger and the Question of Animality." *International Journal of Philosophical Studies* 4, 1 (1996): 67–86.

Goldstein, Kurt. *The Organism: A Holistic Approach to Biology Derived from Pathological Data in Man*. New York: Zone Books, 2000.

Gould, Stephen J. *Ontogeny and Phylogeny*. Cambridge: Harvard University Press, 1977.

Grene, Marjorie. *Approaches to a Philosophical Biology*. New York: Basic Books, 1968.

Haar, Michel. *The Song of the Earth: Heidegger and the Grounds of the History of Being*. Translated by Reginald Lilly. Bloomington: Indiana University Press, 1993.

Hansen, Mark B. N. "The Embryology of the (In)visible." *The Cambridge Companion to Merleau-Ponty*. Edited by Taylor Carman and Mark B. N. Hansen. Cambridge: Cambridge University Press, 2005.

Haraway, Donna J. *Crystals, Fabrics, and Fields: Metaphors of Organicism in Twentieth-Century Developmental Biology*. New Haven: Yale University Press, 1976.

Hardt, Michael. *Gilles Deleuze: An Apprenticeship in Philosophy*. Minneapolis: University of Minnesota Press, 1993.

Harmon, Graham. *Tool-Being: Heidegger and the Metaphysics of Objects*. Chicago: Open Court, 2002.

Heidegger, Martin. GA 2. *Sein und Zeit*. Tübingen: Max Niemeyer Verlag, 1927. (*Being and Time*. Translated by John Macquarrie and Edward Robinson. Oxford: Blackwell, 1962.)

———. GA 8. *Was Heißt Denken?* Tübingen: Max Niemeyer Verlag, 1954. (*What Is Called Thinking?* Translated by J. Glenn Gray. New York: Harper & Row, 1968.)

———. GA 9. *Wegmarken*. Frankfurt am Main: Vittorio Klostermann, 1976. (*Pathmarks*. Edited by William McNeill. Cambridge: Cambridge University Press, 1998.)

———. GA 12. *Unterwegs zur Sprache*. Pfullingen: Verlag Günther Neske, 1959. (*On the Way to Language*. Translated by Peter D. Hertz. New York: HarperCollins, 1982.)

———. GA 15. *Heraklit Seminare* (with Eugen Fink). Frankfurt am Main: Vittorio Klostermann, 1970. (*Heraclitus Seminar*. Translated by Charles H. Seibert. Evanston, IL: Northwestern University Press, 1993.)

———. GA 20. *Prolegomena zur Geschichte des Zeitbegriffs*. Frankfurt am Main: Vittorio Klostermann, 1979. (*History of the Concept of Time: Prolegomena*. Translated by Theodore Kisiel. Bloomington: Indiana University Press, 1992.)

---. GA 21. *Logik: Die Frage nach der Wahrheit*. Frankfurt am Main: Vittorio Klostermann, 1976.

---. GA 24. *Die Grundprobleme der Phänomenologie*. Frankfurt am Main: Vittorio Klostermann, 1975. (*The Basic Problems of Phenomenology*. Translated by Albert Hofstadter. Bloomington: Indiana University Press, 1982.)

---. GA 26. *Metaphysische Anfangsgründe der Logik im Ausgang von Leibniz*. Frankfurt am Main: Vittorio Klostermann, 1978. (*The Metaphysical Foundations of Logic*. Translated by Michael Heim. Bloomington: Indiana University Press, 1984.)

---. GA 29/30. *Die Grundbegriffe der Metaphysik. Welt-Endlichkeit-Einsamkeit*. Frankfurt am Main: Vittorio Klostermann, 1983. (*The Fundamental Concepts of Metaphysics: World, Finitude, Solitude*. Translated by William McNeill and Nicholas Walker. Bloomington: Indiana University Press, 1995.)

---. GA 31. *Vom Wesen der menschlichen Wahrheit*. Frankfurt am Main: Vittorio Klostermann, 1982. (*The Essence of Human Freedom: An Introduction to Philosophy*. Translated by Ted Sadler. London: Continuum Press, 2005.)

---. GA 33. *Aristoteles, Metaphysik Θ 1–3: Von Wesen und Wirklichkeit der Kraft*. Frankfurt am Main: Vittorio Klostermann, 1990. (*Aristotle's Metaphysics Θ 1–3: On the Essence and Actuality of Force*. Translated by Walter Brogan and Peter Warnek. Bloomington: Indiana University Press, 1995.)

---. GA 40. *Einführung in die Metaphysik*. Tübingen: Max Niemeyer Verlag, 1953. (*Introduction to Metaphysics*. Translated by Gregory Fried and Richard Polt. New Haven: Yale University Press, 2000.)

---. GA 42. *Schellings Abhandlung Über das Wesen der menschlichen Freiheit*. Tübingen: Max Niemeyer Verlag, 1971. (*Schelling's Treatise on the Essence of Human Freedom*. Translated by Joan Stambaugh. Athens: Ohio University Press, 1985.)

---. GA 54. *Parmenides*. Frankfurt am Main: Vittorio Klostermann, 1982. (*Parmenides*. Translated by André Schuwer and Richard Rojcewicz. Bloomington: Indiana University Press, 1992.)

---. GA85. *Vom Wesen der Sprache. Die Metaphysik der Sprache und die Wesung des Wortes. Zu Herders Abhandlung "Über den Ursprung der Sprache."* Frankfurt am Main: Vittorio Klostermann, 1999. (*On the Essence of Language: The Metaphysics of Language and the Essencing of the Word. Concerning Herder's Treatise On the Origin of Language*. Translated by Wanda Torres Gregory and Yvonne Unna. Albany, NY: State University of New York Press, 2004.)

---. *Basic Writings*. Edited by David Farrell Krell. New York: HarperCollins, 1993.

---. *Supplements: From the Earliest Essays to* Being and Time *and Beyond*. Edited by John van Buren. Albany: State University of New York Press, 2002.

---. "Vom Wesen der Wahreit," unpublished archival manuscript (photocopy). Lecture delivered Monday, May 24, 1926, in Marburg.

---. *Becoming Heidegger: On the Trail of His Early Occasional Writings, 1910–1927*. Edited by Theodore Kisiel and Thomas Sheehan. Evanston, IL: Northwestern University Press, 2007.

Heusden, Barend van. "Jakob von Uexküll and Ernst Cassirer." *Semiotica* 134, 1/4 (2001): 275–292.

Hoffmeyer, Jesper. *Signs of Meaning in the Universe*. Translated by Barbara J. Haveland. Bloomington: Indiana University Press, 1993.

Hume, David. *Dialogues Concerning Natural Religion*. Amherst, NY: Prometheus Books, 1989.

Jacob, François. *The Logic of Life*. Translated by Betty E. Spillman. New York: Pantheon, 1973.

Jonas, Hans. *The Phenomenon of Life: Toward a Philosophical Biology*. New York: Harper & Row, 1982.

———. *Mortality and Morality: A Search for the Good after Auschwitz*. Edited by Lawrence Vogel. Evanston, IL: Northwestern University Press, 1996.

Kant, Immanuel. *The Critique of Judgement*. Translated by James Creed Meredith. Oxford: Oxford University Press, 1952.

———. *Critique of Pure Reason*. Edited and translated by Paul Guyer and Allen W. Wood. Cambridge: Cambridge University Press, 1999.

Kauffman, Stuart A. *The Origins of Order: Self-Organization and Selection in Evolution*. Oxford: Oxford University Press, 1993.

Kisiel, Theodore. *The Genesis of Heidegger's Being and Time*. Berkeley: University of California Press, 1993.

Klaver, Irene. "Stone Worlds: Phenomenology on (the) Rocks." *Eco-Phenomenology: Back to the Earth Itself*. Edited by Charles S. Brown and Ted Toadvine. Albany: State University of New York Press, 2003.

Kockelmans, Joseph, and Kisiel, Theodore, eds. *Phenomenology and the Natural Sciences*. Evanston, IL: Northwestern University Press, 1970.

Kofman, Sarah. *Comment s'en sortir?* Paris: Galilée, 1983.

Krell, David Farrell. *Daimon Life: Heidegger and Life Philosophy*. Bloomington: Indiana University Press, 1992.

Kuhn, Thomas. *The Structure of Scientific Revolutions*. Chicago: University of Chicago Press, 1962.

Kull, Kalevi. "Biosemiotics in the twentieth century: a view from biology." *Semiotica* 127, 1/4 (1999): 385–414.

———. "Jakob von Uexküll: An introduction." *Semiotica* 134–1/4 (2001): 1–59.

Lacan, Jacques. "The mirror stage as formative of the function of the I as revealed in psychoanalytic experience." *Écrits: A Selection*. Translated by Alan Sheridan. New York: W. W. Norton, 1977.

Lassen, Harald. "Leibniz'sche Gedanken in der Uexküll'schen Umweltlehre." *Acta Biotheoretica* A5 (1939): 41–50.

Lawlor, Leonard. *Thinking Through French Philosophy: The Being of the Question*. Bloomington: Indiana University Press, 2003.

Lenoir, Timothy. *The Strategy of Life: Teleology and Mechanics in Nineteenth Century German Biology*. Chicago: University of Chicago Press, 1982.

Levinas, Emmanuel. "The Paradox of Morality: an Interview with Emmanuel Levinas." Translated by Andrew Benjamin and Tamra Wright. *The Provocation of Levinas: Rethinking the Other*. Edited by Robert Bernasconi and David Wood. New York: Routledge Press, 1988.

Levins, Richard, and Richard Lewontin. *The Dialectical Biologist*. Cambridge: Cambridge University Press, 1985.

Lingis, Alphonso. "Orchids and muscles." *Exceedingly Nietzsche*. Edited by David Farrell Krell and David Wood. New York: Routledge Press, 1988.

———. "Segmented Organisms." *Merleau-Ponty, Interiority and Exteriority, Psychic Life and the World*. Edited by Dorothea Olkowski and James Morley. Albany: State University of New York Press, 1999.

———. "The World as a Whole." *From Phenomenology to Thought, Errancy, and Desire: Essays in Honor of William J. Richardson, S.J.* Edited by Babette Babich. Dordrecht: Kluwer, 1995.

———. "Beastiality." *Animal Others: On Ethics, Ontology, and Animal Life*. Edited by H. Peter Steeves. Albany: State University of New York Press, 1999.

Lippit, Akira Mizuta. *Electric Animal: Toward a Rhetoric of Wildlife*. Minneapolis: University of Minnesota, 2000.

Llewelyn, John. *Seeing Through God: A Geophenomenology*. Bloomington: Indiana University Press, 2003.

Lorenz, Konrad. "Methods of approach to the problems of behavior." *Studies in Animal and Human Behavior (Volume II)*. Translated by Robert Martin. Cambridge: Harvard University Press, 1971.

Lovelock, James. *Gaia: A New Look at Life on Earth*. Oxford: Oxford University Press, 1979.

Low, Douglas. *Merleau-Ponty's Last Vision: A Proposal for the Completion of* The Visible and the Invisible. Evanston, IL: Northwestern University Press, 2000.

———. "The continuing relevance of *The Structure of Behavior*." *International Philosophical Quarterly* 44, 3 (2004): 411–430.

Löwith, Karl. *Nature, History, and Existentialism*. Evanston, IL: Northwestern University Press, 1966.

Mallet, Marie-Louise, ed. *L'animal autobiographique: autour de Jacques Derrida*. Paris: Galilée, 1999.

Maltzahn, Kraft E. von. *Nature as Landscape: Dwelling and Understanding*. Montreal and Kingston: McGill-Queen's University Press, 1994.

Margulis, Lynn. *Symbiotic Planet: A New Look at Evolution*. New York: Basic Books, 1998.

Massumi, Brian. "The Autonomy of Affect." *Deleuze: A Critical Reader*. Edited by Paul Patton. Oxford: Blackwell, 1996.

———. *Parables of the Virtual: Movement, Affect, Sensation*. Durham, NC: Duke University Press, 2002.

Mazis, Glen A. "Merleau-Ponty's Concept of Nature: Passage, The Oneiric, and Interanimality." *Chiasmi International. Merleau-Ponty: From Nature to Ontology*. VRIN, Mimesis, University of Memphis Press, 2000.

McFadden, Johnjoe. *Quantum Evolution: How Physics' Weirdest Theory Explains Life's Biggest Mystery*. New York: W. W. Norton, 2000.

McNeill, William. "Metaphysics, Fundamental Ontology, Metontology 1925–35." *Heidegger Studies* 8 (1992): 63–79.

———. *The Glance of the Eye: Heidegger, Aristotle, and the Ends of Theory*. Albany: State University of New York Press, 1999.

———. "Life Beyond the Organism: Animal Being in Heidegger's Freiburg Lectures, 1929–30." *Animal Others: On Ethics, Ontology, and Animal Life*. Edited by H. Peter Steeves. Albany: State University of New York Press, 1999.

———. *The Time of Life: Heidegger and Êthos.* Albany: State University of New York, 2006.

Merleau-Ponty, Maurice. *La structure de la comportement.* Paris: Presses Universitaires de France, 1943). (*The Structure of Behavior.* Translated by Alden Fischer. Pittsburgh: Duquesne University Press, 1983).

———. *Phénoménologie de la perception.* Paris: Gallimard, 1945. (*The Phenomenology of Perception.* Translated by Colin Smith. New York: Routledge Press, 1998).

———. *Signes.* Paris: Gallimard, 1960. (*Signs.* Translated by Richard C. McCleary. Evanston, IL: Northwestern University Press, 1964).

———. *Le visible et l'invisible.* Paris: Gallimard, 1963. (*The Visible and the Invisible.* Translated by Alphonso Lingis. Evanston, IL: Northwestern University Press, 1968).

———. *La Nature: Notes Cours du Collége de France,* et. Dominique Séglard. Paris: Seuil, 1995. (*Nature: Course Notes from the Collége de France.* Translated by Robert Vallier. Evanston, IL: Northwestern University Press, 2003).

———. *Notes des cours au Collège de France: 1957–58 et 1960–61.* Paris: Gallimard, 1996.

Monod, Jacques. *Chance and Necessity: An Essay on the Natural Philosophy of Modern Biology.* New York: Knopf, 1971.

Morris, David. "Animals and humans, thinking and nature." *Phenomenology and the Cognitive Sciences,* 4 (2005): 49–72.

Murphy, Tim. "Quantum ontology: A virtual mechanics of becoming." *Deleuze and Guattari: New Mappings in Politics, Philosophy, and Culture.* Edited by Eleanor Kaufman and Kevin Jon Keller. Minneapolis: University of Minnesota Press, 1998.

Nagel, Thomas. "What is it like to be a bat?" *The Philosophical Review* LXXXIII, 4 (1974): 435–50.

Ortega y Gasset, José. "Preface" to Jakob von Uexküll, *Ideas para una concepción biológica del mundo.* Translated by R. M. Terneiro. Madrid: Calpe, 1934.

Plato. *The Republic, Books I–V.* Translated by Paul Shorey. Cambridge: Harvard University Press, 1969.

———. *Protagoras.* Translated by W. R. M. Lamb. Cambridge: Harvard University Press, 1999.

———. *Phaedo.* Translated by Harold North Fowler. Cambridge: Harvard University Press, 1999.

Richir, Marc. *Phenomenologie et Institution Symbolique.* Paris: Éditions Jerome Millon, 1988.

Russon, John. "Embodiment and responsibility: Merleau-Ponty and the ontology of nature." *Man and World* 27 (1994): 291–308.

Safranski, Rüdiger. *Martin Heidegger: Between Good and Evil.* Translated by Ewald Osers. Cambridge: Harvard University Press, 1998.

Salthe, Stanley N. *Development and Evolution: Complexity and Change in Biology.* Cambridge: MIT Press, 1993.

Sartre, Jean-Paul. *Being and Nothingness.* Translated by Hazel E. Barnes. New York: Gramercy Books, 1956.

Schalow, Frank. "Who Speaks for the Animals? Heidegger and the Question of Animal Welfare." *Environmental Ethics* 22 (2000): 259–271.

———. *The Incarnality of Being: The Earth, Animals, and the Body in Heidegger Thought*. Albany: State University of New York Press, 2006.
Scheler, Max. *Die Stellung des Menshen im Kosmos*. In *Max Scheler, Gesammelte Werke (Band 9): Späte Schriften*. Bern: Francke Verlag, 1976. (*Man's Place in Nature*. Translated by Hans Meyerhoff. Boston: Beacon Press, 1961.)
Schiller, Claire, ed. *Instinctive Behavior*. New York: International Universities Press, 1957.
Sebeok, Thomas A. *Perspectives in Zoosemiotics*. The Hague: Monton, 1972.
———. *Signs: An Introduction to Semiotics*. Toronto: University of Toronto Press, 1994.
———. *Global Semiotics*. Bloomington: Indiana University Press, 2001.
Smolin, Lee. *The Life of the Cosmos*. Oxford: Oxford University Press, 1997.
Steeves, H. Peter, ed. *Animal Others: On Ethics, Ontology, and Animal Life*. Albany: State University of New York Press, 1999.
———. "The familiar other and feral selves: Life at the human/animal boundary." *The Animal/Human Boundary: Historical Perspectives*. Edited by Angela Creager and William Chester Jordan. Rochester, NY: University of Rochester Press, 2002.
———. "A Quantum Magical Realism Writ Small: Self-referentiality, information, and the origin of life." Unpublished manuscript, 2005.
———. *The Things Themselves: Phenomenology and the Return to the Everyday*. Albany: State University of New York, 2006.
Thomson, T. Arthur. Review of *Theoretical Biology*, by Jakob von Uexküll. *Journal of Philosophical Studies* 2(7), (1927): 413–419.
Toadvine, Ted. "Singing the World in a New Key: Merleau-Ponty and the Ontology of Sense." *Janus Head* 7, 2 (2004): 273–283.
Uexküll, Jakob von. *Umwelt und Innenwelt der Tiere*. Berlin: J. Springer, 1909/1921. ("Environment and Inner World of Animals." Translated excerpts by Chauncey J. Mellor and Doris Gove. *Foundations of Comparative Ethology*. Edited by Gordon M. Burghardt. New York: Van Nostrand Reinhold, 1985).
———. "Kant als Naturforscher. Von Erich Adikes." *Deutsche Rundschau* 53, 5 (1924): 209–210.
———. *Theoretische Biologie*. Berlin: J. Springer, 1926. (*Theoretical Biology*. Translated by D. L. Mackinnon. New York: Harcourt, Brace, 1927).
———. *Niegeschaute Welten*. Berlin: S. Fischer Verlag, 1936. ("An introduction to Umwelt." Translated excerpts by Gösta Brunow. *Semiotica* 134, 1/4 (2001): 107–110.)
———. "Kants Einfluß auf die heutige Wissenschaft: Der große Königsberger Philosoph ist in der Biologie wieder lebendig geworden." *Preußische Zeitung* 9, 43 (1939): 3.
———. "The new concept of Umwelt: A link between science and the humanities." Translated by Gösta Brunow. *Semiotica* 134, 1/4 (2001): 111–123.
———. *Bedeutungslehre*. Leibzig: Verlag von J. A. Barth, 1940. ("The theory of meaning." *Semiotica* 42, 1 (1982): 25–82).
———. *Mondes Animaux et Monde Humain, suivi de Théorie de la Signification*. Hamburg: Éditions Gonthier, 1956.

Uexküll, Jakob von, and Kriszat, Georg. *Streifzüge durch die Umwelten von Tieren und Menschen*. Geibungsyoin Verlag, 1934. ("A Stroll Through the Worlds of Animals and Men." *Instinctive Behavior*. Edited and translated by Claire Schiller. New York: International Universities Press, 1957).

Uexküll, Gudrun von. *Jakob von Uexküll: Seine Leben und seine Umwelt, Eine Biographie*. Hamburg: Christian Wegner Verlag, 1964.

Uexküll, Thure von. "Introduction: The Sign Theory of Jakob von Uexküll." *Semiotica* 89, 4 (1992): 279–315.

Villani, Arnaud. *La guêpe et l'orchidée: Essai sur Gilles Deleuze*. Paris: Belin, 1999.

Žižek, Slavoj. *Organs Without Bodies: On Deleuze and Consequences*. New York: Routledge Press, 2004.

Index

absence, 80–83, 85–86; invisibility, 143–44. *See also* presence
absorption, 61, 77–78, 87, 195
abyss, 53, 67, 79, 99–100, 105–6, 113, 148, 196
actual, the, 161–62, 165–67, 169, 172
affects, 36, 93; as becoming, 156–60, 182–85, 190; body, 6, 158–61, 166, 172–73, 183–86, 190; the counting of, 6, 156–59, 161, 184, 186, 190; ethology of, 154–62, 182, 186, 190; Spinozian, 158, 190; study of, 155–58; the tick's three, 155–56, 177
Agamben, Giorgio, 3, 12, 78, 12, 87, 90; *The Open*, 78
ambiance. *See* milieu
amoeba, 28, 35, 67, 71, 75, 79
animal: animal-stalks-at-five-o'clock, 182–86; becoming-animal, 38, 182, 183; behavior, 3, 5, 7–8, 21, 37, 50, 76–97, 109, 113, 117, 121, 133–35, 159, 186, 188, 189, 196, 199; being, 2, 4–6, 45, 38, 48–49, 66–68, 73, 77, 84, 93, 106–8, 111, 118, 121, 147, 189–90; body, 3–6, 37–38, 50–51, 95–96, 113, 144; capacity, 75–78, 83, 95, 197; concept of, 68; and environment, 47–48, 51–54, 73, 97, 107, 143; essence of, 44, 45, 48, 67–68, 73–74, 90, 106; identity of an, 29; interanimality, 135, 139, 145–49; language, 30–33, 98, 101–3, 113; meaning of being an, 2, 43; melody, 6, 28, 115, 124–30, 136–38; as non-

transcendent, 77, 102–6; openness of the, 88–89, 94, 98–99; as poor in world, 38, 43, 63, 67, 71, 73–74, 93, 99, 112–14, 189; self, 75–77, 95, 109, 146, 148; self-encirclement, 93–96, 98, 103; as subjective, 2–3, 7, 13, 21–22, 37, 133, 187; symphony, 26–28, 124–25; as taken, 78, 89, 103, 109; time, in relation to, 47, 63, 92, 99–111, 165, 177; totality of the, 6, 51, 77, 93, 95–97, 122–23, 127, 135, 138; world, as having, 93
Ansell Pearson, Keith, 2, 3, 154, 184, 191, 202, 206
anthropomorphic, 8, 20
Aquinas, Saint Thomas, 163
archeology, 118, 198
Aristotle, 35, 79, 195, 202; *Historia Animalium*, 79; potentiality, 35
Artaud, Antonin, 151
assemblages, 38, 171–74, 184–86; territorialized, 178. *See also* strata
'as' structure, 69, 72, 100–1, 105
atomism, 19, 118–22, 126, 161
Augustine, Saint, 56–57
authenticity, 60, 75, 76, 107

Badiou, Alain, 152, 162, 182, 203
Baer, Karl Ernst von, 8–12, 14, 31, 39, 44, 46–47, 54, 154, 167; Baer's law, 10, 192; *Über Darwins Lehre*, 11
becoming, 153, 156, 158–60, 165–69, 170, 172, 177, 184–85, 203; animal, 6, 38, 154, 179, 182–83, 190;

becoming (continued)
  genetic, 182; molecular, 173; other, 34, 161; rhizomatic, 180; of the world, 189. See also affects
bee, 1, 5, 26, 43, 67, 79–89 passim, 98, 196–97; and flower, 33–36, 179–80
behavior, 4, 21, 36–37, 60, 78, 81–89, 93–96, 98–99, 101, 109, 115–22, 125, 133–35, 143–44, 155, 186, 188–89, 190, 192, 199; animal vs. human, 38, 76–79, 91, 97, 113–14. See also ethology
behaviorism, 119, 133
being: analogical, 162–63; being-in-itself, 70; brute or wild, 115, 139–40, 147; equivocal, 163–64; genetic, 167, 173–74; the great chain of, 64, 162; hollow, 139–40, 144, 148; interbeing, 147; leaf of, 138–44; meaning of, 41–42, 131; naïve, 145; natural, 139–40; primordial, 101, 105–6; question of, 40, 152; restlessness, 84; sensible, 140–48; temporal dimension of, 101; univocal, 162–64, 172, 183–85; whole or totality of, 59, 84–85, 97, 147
being-in-the-world, 38, 44, 55, 59, 62, 102, 121–22, 183
beings: individuation of, 163–66, 173–74, 185
Benjamin, Walter, 12
Bergson, Henri, 135; Creative Evolution, 29
biogenetics, 176
biology, 7, 9–11, 20, 28, 46, 54, 96, 118, 154, 205; Deleuze and, 167, 171, 182; Driesch's, 49–51; Heidegger and, 5, 39, 40–46, 51–53, 55, 64, 87, 92, 94, 96, 109; history of, 44, 52; Merleau-Ponty and, 116, 118–19, 133, 137, 147; vs. ontology, 41, 68; as a theory of life, 30; Uexküll's, 2, 4–5, 8, 10–16, 19, 21, 31, 37–38, 53, 55, 96, 109, 116, 123, 137, 147, 149, 154
biophilosophy, 3, 4, 154

biosemiotics, 5, 8, 28–36; sign systems, 5, 31
body, 3, 4, 19, 34, 65, 107, 112, 123, 151–53, 159, 161, 167, 201; affects, 6, 158–61, 166, 172–73, 177, 183–86, 190; becoming-other, 161; as chiasm, 130, 141, 142; Dasein, 49; human, 97, 108, 114, 147; individuation of the, 29–30, 153, 160, 164, 183; limit of the, 95, 133; moving, 131, 148; phantom limb, 82; porous, 161, 174–75; problem of the, 114–15; ring, 94; role of the, 3, 12, 37–38; as singing, 136–42; skin, 140, 175; spatiality of the, 135, 140; surface, 51, 95–96; as a two-in-one, 142; and umwelt, 6, 37, 115, 143–46, 181, 189; as a whole, 60, 96; world, cohesion with, 38, 50, 132–33, 147. See also lived body
body without organs, 161–62, 169–74, 185
Boveri, Theodor: The Organism as a Historical Being, 111
Brentano, Franz, 35
Bussoti, Sylvano, 27
Buytendijk, F. J. J., 44, 49–52, 123, 128, 44, 49, 51–52, 95–96; Investigations on the Essential Differences between Humans and Animals, 50

Canguilhem, Georges, 3, 7, 191, 199
captivation, 5, 73–74, 78, 83, 89, 103, 106, 109, 111–12; behavior, as form of, 76; vs. Dasein, 105; essence of, 75, 99; structure of, 77; unity of, 50, 95–97
Carroll, Lewis, 161, 168
Cassirer, Ernst, 3, 31, 40, 191, 193; Philosophy of Symbolic Forms, 31
chaos, 27, 170, 174, 180–81
chaosmos, 174, 204–5
chiasm, 132–33, 141–47, 181. See also flesh
circle, 92–96, 104, 112, 144–46, 180–81; animal, 38, 88–110 passim;

concentric, 145, 176; cracked, 99, 181–82, 190; of the philosopher, 90
complexity theory, 153, 174, 181
comportment, 38, 76, 83, 87, 102–3, 111, 197; vs. animal behavior, 5, 76–78, 91, 113, 189; vs. *comportement*, 76, 199
consciousness, 3, 5, 29, 35, 66, 70, 117, 119, 121, 129–31, 134, 139–40, 143, 187; self, 42, 75
coordination, 120, 122, 126–27
Cuvier, Georges, 10, 154

Darwin, Charles, 8–11, 17–20, 31, 44, 47–48, 50, 52, 94, 154, 168, 171, 192, 203–4; Darwinism, 16–19, 46–48, 192, 204; morphology, 11, 47; natural selection 8, 10–11, 47–48, 203; *The Origin of Species*, 10, 203. *See also* evolution
Dasein: absorption, 58, 61, 96, 127, 186, 195; vs. animals, 5, 38, 43–45, 49, 53, 75–76, 84–85, 91, 98–100, 110–13, 188–89; and the body, 49, 107; existential analytic of, 39–41; vs. life, 41–42, 66–67, 197; temporality of, 104–8, 111; transcendence of, 101–3; world of, 43, 53–56, 59, 62–64, 193
Dennett, Daniel, 11, 35, 47, 171; *Darwin's Dangerous Idea*, 47
De Landa, Manuel, 154, 156, 168, 170, 202–4; *Intensive Science and Virtual Philosophy*, 154, 202
Deleuze, Gilles, 3, 4, 6, 8, 9, 24, 27, 30, 33, 38, 71, 74, 75, 83, 84, 93, 100, 110, 112, 113, 119, 123, 127, 135, 139, 149, 151–86 *passim*, 187–92, 199–200; *Difference and Repetition*, 6, 151, 154, 162, 165; *Expressionism in Philosophy: Spinoza*, 158; *Logic of Sense, The*, 154, 161, 168, 201; *Spinoza: Practical Philosophy*, 154–55, 157, 187, 201–2
Deleuze and Guattari; 3, 9, 27, 38, 83, 93, 100, 127, 192, A Thousand

Plateaus, 27, 151, 155, 169, 170, 173, 176, 177, 180, 182, 203;
What is Philosophy?, 155, 177, 188, 192
Derrida, Jacques, 67–68, 72, 106, 113, 195, 196, 201; *Of Spirit*, 67, 72; *The Animal that Therefore I Am (More to Follow)*, 67, 198
Descartes, Rene, 58, 84; Cartesian metaphysics, 133
deterritorialization. *See* territories: de- and reterritorialization
difference, 163–65; differentiation; 159, 165–66; differenciation; 169, 171
Dilthey, Wilhelm, 40, 92
disinhibition, 93–94, 97, 99, 106
disruption, 82, 99, 101, 106
Driesch, Hans, 44, 50–52, 96, 148; 1907–1908 Gifford Lectures, 51
duet, 26, 28–30; of bee and flower; 33, 80, 180

egg, 168–71: embryo, 10, 50, 134, 166; embryology, 153, 167–68, 190; world as an, 169
embodiment. *See* body
entropy, 174, 204
environment, 1–3, 7–8, 21–23, 39–40, 50–55, 66, 73, 91, 95, 112–14, 124–30, 133, 146, 188–90, 199
epigenesis, 10, 18–19
equipment, 59–60; totality of, 61. *See also* organs
ethology, 2, 21, 137, 155, 159–60, 186, 189–90, 205; onto-ethology, 4, 135, 184–85, 188, 191–92; Spinozian ethologist, 6, 155, 157–58, 190; as study of affects, 156–58, 186. *See also* behavior
everydayness, 5, 54, 56, 59–62
evolution, 11, 47–48, 97; aparallel, 179; Darwinian, 8–11, 17–20, 31, 192; intentionality, 35
existence, 38, 40–42, 53, 64, 70, 115
extensive, 157, 161, 165, 172–74, 202, 203

finitude. See time
flesh, 50, 116, 130–48 passim, 181, 200. See also chiasm
folds: of being, 139–42, 144, 147; evolutionary, 18
form, 119, 129. See also Gestalt
Foucault, Michel, 4, 194, 198, 201, 202, 203
Freud, 80, 203
Frisch, Karl von, 79

Gadamer, Hans-Georg, 3
Gestalt, 85, 184; psychology, 122; theory, 119, 122
geophenomenology, 4
geophilosophy, 4
God, 10, 57, 64, 113, 163–64, 170, 202; absence of, 16–18, 56
Goethe, Johann Wolfgang von, 33
Goldstein, Kurt, 122–23, 125, 142, 199
Greek: concept of world, 57; philosophy, 40
Grene, Marjorie, 3, 8
Guattari, Félix. See Deleuze and Guattari

haecceity, 183–84
hands, 79, 97, 141, 143, 196–97
Haraway, Donna, 45
Hegel, G.W.F., 9, 30, 193
Heidegger, Martin, 3–5, 8–9, 22–23, 30, 39–114 passim, 115–18, 120–23, 127–28, 132, 136, 145–49, 152, 154, 162, 165–66, 175–77, 181, 183, 185, 188–91, 199–204; *1926 lecture in Marburg*, 92; *The Basic Problems of Phenomenology*, 104, 193; *Being and Time*, 39–43, 46, 49, 54, 56, 58, 59, 62, 66, 75, 92, 93, 96, 101, 102, 105–7, 110, 194; *The Fundamental Concept of Metaphysics*, 41, 43, 45, 46, 55, 56, 59, 61, 62, 64, 66, 67, 73, 91, 92, 93, 97, 98, 100–2, 106–8, 112, 114, 123, 196, 199, 204; *Heraclitus Seminar*, 52, 69, 114; *Introduction to Metaphysics*, 73; "Letter on 'Humanism,'" 97, 98, 102, 109, 112; *Logic*, 46; *Metaphysical Foundations of Logic*, 56, 58, 102; "On the Essence and Concept of Phusis in Aristotle's Physics B, 1," 63, 74; "On the Essence of Ground," 56, 102; *On the Origin of Language*, 52; "On the Question of Being," 72; "The Origin of the Work of Art," 63; *Parmenides*, 87; *What is Metaphysics?*, 64, 103; "Wilhelm Dilthey's Research and the Struggle for a Historical Worldview," 92
Helmholtz, Hermann von, 13–15
Heraclitus, 52, 56, 69, 114, 188
Herder, Johann Gottfried, 104; *On the Origin of Language*, 52
home, 52, 83–86, 180–81; being at, 84, 181; homesickness, 83–85
human: vs. animals, 48–49, 53, 67–68, 97, 101–2, 112; defining the, 42; as ek-sisting, 97; essence of, 67, 74; existence, 40–41, 64; as ontic, 40, 165; as a whole, 122; as world forming, 43, 63, 67, 102. See also body: human
Hume, David: *Dialogues Concerning Natural Religion*, 17, 200
Husserl, Edmund, 29, 35, 40

inorganic forces, 21, 36
interanimality, 6, 135–36, 139, 145, 146
intentionality, 29; body, 35
intensive, 6, 161–62, 165, 172–73, 178, 182–84, 203; forces, 174; intensities, 157, 161, 167, 169, 203
intercorporeity, 141. See also body

Jonas, Hans, 3, 49, 66, 113; "Philosophy at the End of the Century," 49

Kant, Immanuel, 8–10, 12–16, 20–21, 31, 37, 42, 55–58, 153, 192; *1770 dissertation*, 57; *Critique of Pure Reason*, 13, 57–58; *Logic*, 42
Koffka, Kurt, 122

Köhler, Wolfgang, 122
Kühne, Wilhelm, 12
Kull, Kalevi, 18, 31

Lacan, Jacques, 3
Lamarck, Jean-Baptiste, 9
language, 100, 112; 'as' of, 100. See also animal: language
Lefort, Claude, 138
Leibniz, Gottfried, 23, 58, 200
Levinas, Emmanuel, 48–49
life, 12, 44–45, 76, 130, 148–49, 160–61; vs. existence, 42, 67; lifeless things, 36; origin of, 149, 171, 173; as process, 111–12; struggle for, 49; and world, 92–93
light, 155–57, 159
lines of flight, 38, 93, 178–82, 190, 203
lived body, 116, 128, 131–32, 138, 140–45, 160
Lorenz, Konrad, 2, 8, 36–37, 196
Löwith, Karl, 41, 56–57, 66, 113
Lucretius, 157

Margulis, Lynn, 33
materialism, 127; materialist monism 65; ontological materialism, 161; vital, 10, 14
meaning, 8, 12–13, 15, 30, 37, 128, 129, 134, 141; acquisition of, 35; generation of, 32, relations of, 30; unity of, 127
mechanism, 9–11, 16, 120, 201. See also vitalism: and mechanism
melody, 124–30, 136–38, 142–43; fly's, 34; self singing, 123–24, 129–30, 136–37
Merleau-Ponty, Maurice, 3–6, 8, 9, 23, 30, 38, 50, 60, 65, 71, 76, 77, 82, 111, 114, 115–50 passim, 152, 154, 162, 167, 175, 177, 181, 185, 188–90; *Nature*, 117, 123, 133, 137, 138, 148; *Phenomenology of Perception*, 117, 131, 134, 140, 188, 200; *The Structure of Behavior*, 50, 76, 116–18, 122–23, 125, 128, 130–31, 133, 144, 147, 188, 199; *The Visible and the Invisible*, 50, 115–17, 131, 138, 152, 200
milieu, 123, 127, 135–37, 174–76; coded, 176–78; field of action, 135–36, 148; vs. territories, 178. See also environment
morphology, 47, 88; animal, 95–96; development, 95; embryonic, 10; role of, 14; structure, 94, 96; teleological view of, 11
motion, 111–12, 190
movement, 134–35, 166, 173, 189; body, 131, 142; of particles, 161, 184; unity of, 134–35
multiplicity, 164–66
music, 176–77, 181; instruments of, 124–25; of life, 26–29. See also melody

nature: as anthropomorphized, 20; horizontal vs. vertical view of, 20; intersubjective theory of, 28–29; as music, 25–29, 124, 177, 180–81; ontology of, 115–17, 129–33, 138–39, 147, 189; plane of, 161, 170, 186, 190; as web of life, 20–21, 25, 34; vs. world, 56–57; See also plan of nature
naturalism, 117–18
Naturphilosophie, 9, 137
Newton, Isaac, 9, 16–17; Newtonian physics, 20, 153
Nietzsche, Friedrich, 15, 55, 117, 152, 162
Novalis, 83

ontic, 62, 165; animal as, 74, human as, 40; proximity, 69–70. See also ontological difference, the
ontogeny, 11, 184; ontogenesis, 148. See also phylogeny
onto-hetero-genesis, 165
ontological difference, the, 41, 162, 165, 201; humans and animals, 76–79, 112, 189; time as the horizon for, 102, 110

ontology: biological, 13, 37; fundamental, 40–43, 152, 165; of life, 42; of nature, 115, 138, 147; return to, 115–16, 131–33, 142, symbiotic, 143
orchid and wasp, 177–80, 205
organicism, 45
organs, 75, 109–10; vs. equipment, 110, 196; vs. instruments, 109; *See also* equipment
Ortega y Gasset, José, 3
organism, 26, 74, 123–24, 134, 149, 151; autonomy, 15; becoming other, 33–36, boundaries of the, 29; capacity, 75; embodied anticipation, 35; as the enemy, 151; essence of the, 46–47; as historical, 111; melody that sings itself, 123–24, 129–30; model, 68; as problem and solution, 151–54, 161; and self, 75–76, 148; as self generative, 74–75; as sign and interpreter, 32–33; whole or totality, 50–51, 119–20, 123–27, 134

Parmenides, 87, 162
Pavlov, Ivan, 119
phenomenology, 35, 116–17, 131, 139; of the body, 50, 82 (*see also* lived body); ecophenomenology, 4; geophenomenology, 4; of perception, 128, 139
phusis, 56, 195
phylogeny, 184. *See also* ontogeny
plane of immanence, 160–61, 168, 170; as musical, 177
plan of nature, 4, 8, 12, 16–21, 27–28; Baer, 10; musical, 30
Plato, 12, 84, 186, 197
plants, 12, 16, 19, 43, 55, 62–64, 67, 160, 162, 195
Plessner, Helmuth: *Die Stufen des Organischen und der Mensch*, 66
Popper, Karl, 37
poverty, 95, 98–99. *See also* animal: as poor in world
prebiotic soup, 171–74
preformationism, theory of, 10, 18

presence, 107–9; absence as, 138; coming-to-presence, 166. *See also* absence
present-at-hand, 40–42, 53, 57, 61, 122, 165; animal as, 47
Proust, Marcel, 136
projection, 100–1

quantum, 171; mechanics, 153, 157; ontology, 161

Radl, Emanuel: *Investigations on Animal Phototropism*, 79
ready-to-hand, 60–62, 107, 165–66
reality, 2–3, 137; objective, 15; structural, 122; as subjective appearance, 13–14, 16, 21–22, 37, 187
refrain, 176–77, 180–82
repetition, 171, 176–78
reterritorialization. *See* territories: de- and reterritorialization
rhizome, 155, 170, 179–80
rhythm, 28, 129–30, 160, 168, 174–81, 204; cellular, 26; circadian, 106
Rilke, Rainer Maria, 88
rings: body, as connected to, 94; disinhibiting, 90, 93–96; encircling, 92–110 *passim*, 181; of finality, 136, 145, 175, 181; limit, 104; shattering of the, 99–100, 175–76

Saint Hilaire, Geoffroy, 154
Sartre, Jean-Paul, 29, 70–71, 85–86; *Being and Nothingness*, 70, 86
Scheler, Max, 40, 66, 91; *Man's Place in Nature*, 91, 66
Schelling, Friedrich, 9, 55
science, 1, 44, 128, 202; cognitive, 65; life, 41; modern, 16, 18, 153–54; and philosophy, 4, 117–18, 147. *See also* biology
Scotus, Duns, 162, 183
seeing, 80, 87, 107; moment of vision, 45, 105, 107; reversibility of, 141–45
Skinner, B. F., 119
snail, the, 91–93, 197

song, 180–82; bird, 180
space, 23, 96, 143; Cartesian, 140, 157; Euclidean, 140, 153; fragmented, 153; Kantian, 20; and time, 11, 20, 47, 153, 165, 168, 176–77; topological, 140–41, 144
sphere. *See* umwelt
spider and fly, 34–35, 177–78
Spinoza, Baruch, 155, 157–58, 162, 187
Stoics, 168, 204
stone, 76, 88, 92, 114, 196; rock, 43, 64–66, 70–71, 98; rock, 72; worldless, 43, 63, 68–72
strata, 127, 170–72, 202; epistrata, 175; shattering of the, 176, 182. *See also* assemblages
structure; 93–97, 118–19, 122–23, 127–30. *See also* being-in-the-world
subject: vs. object, 22, 103, 131, 141, 146, 193
sun, 24, 33, 72, 81, 86–89, 98, 160
symbiosis, 33; symbiotic ontology, 143; symbiotic relationship, 30, 33, 160, 179, 205
symphony, 28, 124–25; of the organism, 26

teleology 4, 9–11, 17, 20; force, 51, 168
territories, 148–49, 174, 177–82; de- and reterritorialization, 178–79, 181–82, 203. *See also* lines of flight
thermodynamics, 157, 204; second law of, 174
things in themselves, 5, 58
tick, 24–25, 32, 36, 94, 98, 154–59, 166, 178, 181, 185–86, 188; and god, 113, 164, 202
time: Aristotelian, 153; ecstatic, 103–11; essence of, 102; finitude, 57–58, 115; and transcendence, 101–6
Tinbergen, Niko, 8

touch, 69, 91–92, 106; reversibility of, 141–43, 146

Uexküll, Jakob von, 1–38 *passim*, 44–46, 50–55, 59, 62, 65, 73, 79, 80, 82, 88, 90, 92–96, 99, 104, 109, 112–16, 122–26, 128–29, 133–34, 137–38, 140, 142–43, 146–49, 151, 154–57, 159, 167, 174–77, 179–81, 185–90, 195, 199, 200; *Die Lebenslehre*, 12; *Environment and Inner World of Animals*, 7, 12, 21, 92; *Niegeschaute Welten*, 12; *A Stroll Through the Environments of Animals and Humans*, 1, 12, 155; *Theoretical Biology*, 7, 12, 14, 16, 18, 29, 79; *The Theory of Meaning*, 12, 30, 179–80
umwelt: 2, 7, 21–22, 125, 134, 137–38, 177; creation, 32; limits, 23–25, 73, 115; overlapping, 25, 28, 145; shattering, 175–76; soap bubble, 1, 23, 25, 38, 90, 126, 175; umweltforschung, 21

virtual, the, 161–62, 167–71, 203; vs. being, 165
vitalism, 118, 127; and mechanism; 45, 118–19; neovitalism, 31, 51, 52; vital force, 127

world: biologists' vs. physicists', 22; concept of, 43, 54–59, 63–64, 121–22, 185; everyday, 59–62; external, 24; vs. life, 92–93; mental, 65–66; *mundus*, 57; sensible, 140–47; subjective, 15–16, 22; totality of, 57–61; worldhood of, 62. *See also* umwelt

zoontology, 4
zoosemiotics. *See* biosemiotics

www.ingramcontent.com/pod-product-compliance
Lightning Source LLC
Chambersburg PA
CBHW020651230426
43665CB00008B/395